U0581230

Club and Lounge

图书在版编目（ＣＩＰ）数据

名流场——俱乐部与休闲吧 / 李春梅编译. --
沈阳：辽宁科学技术出版社，2011.5
　　ISBN 978-7-5381-6597-5

　　Ⅰ．①名… Ⅱ．①李… Ⅲ．①俱乐部－建筑设计
Ⅳ．①TU242.4

中国版本图书馆CIP数据核字（2011）第035783号

出版发行：辽宁科学技术出版社
　　　　　（地址：沈阳市和平区十一纬路29号　邮编：110003）
印 刷 者：利丰雅高印刷（深圳）有限公司
经 销 者：各地新华书店
幅面尺寸：230mm×290mm
印　　张：34
插　　页：4
字　　数：50 千字
印　　数：1~2000
出版时间：2011年5月第1版
印刷时间：2011年5月第1次印刷
责任编辑：陈慈良
封面设计：迟　海
版式设计：迟　海
责任校对：周　文

书　　号：ISBN 978-7-5381-6597-5
定　　价：268.00元

联系电话：024-23284360
邮购热线：024-23284502
E-mail: lnkjc@126.com
http://www.lnkj.com.cn
本书网址：www.lnkj.cn/uri.sh/6597

Club and Lounge

名流场——俱乐部与休闲吧

李春梅 编／译

辽宁科学技术出版社

Club Design
– By Paul Heretakis

Night clubs and Karaoke rooms are generally pretty easy to design because the customers and their expectations are very clearly defined. The energy level is high from the moment the spaces open based upon their limited hours of operations. The customers are also of a certain age group and common interests, so they are all looking for the same form of entertainment. It is all about fantasy and non-stop fun. The design needs to be very overblown and stimulating in order to compete with the music and the visual impact of the people. Multiple textures, colours, sparkling and reflective surfaces are a must. Individually it might look like a collage of insanity, but as a whole it comes together and creates the basis for a great evening. These spaces convey one feeling and one feeling only – an all-out party!

When you design a Flex Club that appeals to older patrons (40 to 60 years old), the expectations and experiences can vary greatly. Due to the economies of space and project costs, utilising a space for more than one use is very beneficial. Many of these spaces start out as restaurants and turn into late nightclubs after 11 pm. As a restaurant, an ocean-front view can create a must-see romantic dining experience. The views become introverted and the exterior windows are closed off since the goal of a nightclub is the voyeuristic element

to the crowd. Being in the moment is the most important factor. Since these spaces Flex uses throughout the day, it gets very difficult to control energy, lighting and mood throughout the day. While a nightclub is generally closed in space with no windows, it can still be a destination and location that is not as important, obviously not the case with a great restaurant. So many of them start out subdued and romantic and build energy as the night moves along. The music or entertainment programme also starts out slower and builds momentum as the night moves on. These clubs also close much earlier. Nightclubs have DJs while the Flex Clubs in the picture will have dueling pianos or live bands. They can be very interactive and entertaining but not nearly as hard driving as the DJ Club can be.

While the fabrics and colours used in a nightclub can be dark and very industrial to hide abuse and damage that nightclub patrons can cause, as many people will dance on a booth or table, and high heels will quickly punch holes in most fabrics or break glass tables. Every material needs be considered very carefully for its durability and look. When you design a Flex Club, the materials need to look new and rich during its restaurant use yet still durable enough for its night-time club use. During restaurant operations the lights would be bright and the materials will appear in a

certain way. Once the transition to the nightclub begins, the lights go down and the materials will lose their lightness and in many cases their luster. Mature and sophisticated colours schemes need to be incorporated in order for the restaurant to compete with other 5-star dining offers. The materials must be as visually flexible as the club itself. The reality or repair and replacement budgets need to be addressed, otherwise the space will appear worn rather quickly.

The Flex Club is a chameleon and always on the move. All of the restaurant operational needs will disappear while the nightclub elements will appear. Dining tables are wheeled out, bar stools are pulled from the bar and replaced with entertainment elements. A large storage area is required as well as areas for dancing, lighting rigs and an extensive sound system, all of which must "disappear" during the day. These elements have lighter duty in a Flex Club, but none the less when the customer is expecting a nightlife experience it must be delivered flawlessly.

These clubs also get rented to MICE as well as hotel specialty events. The opportunities for increased revenue are endless. The Flex Club can become the calling card for the property because of its vast appeal and flexibility.

俱乐部设计
—— 保罗·哈里塔基斯

The Hilton Players Club is a great example of a Flex Club that has built a great dining reputation as well as a cornerstone to nightlife for the property. The Chocolate, blue, green and beige colour scheme is both sophisticated by day and darkens down well in the evening. The materials are very plush where the customer can touch them, and very durable when needed. The walls tend to be lighter so the club lights will shine true on them. The decorative light fixtures are very dramatic and overscaled and work well to fill the large volume of the space. The back lite onyx bar is slightly hidden by the bar stools but comes alive when the stools are removed in the night. Wooden floors throughout lend to various flexible locations for the dance floor. Lounge area for piano entertainment works well for the lighter side of musical entertainment. When a band or DJ plays, the focus is on the dance floor. The dining booths are converted to VIP booths that are rented during the nightclub functions. These tables require bottle minimums and bring in the greatest revenue potential. They're often rented months in advance. The operator must be flexible in their approach and creative with solutions for uses. Of course an entrepreneurial personality will ensure the Flex Club is in constant use.

夜总会和卡拉OK的房间很容易设计，因为顾客和他们的期望很明确。从空间开放的那一刻开始，由于营业时间的限制，通常里面气氛都很热烈。顾客也是特定年龄的人群，而且有着共同的兴趣爱好，所以他们寻求娱乐的方式是相同的。这其中包含了无限的美妙和乐趣。设计需要非常夸张和刺激，以匹配所播放的音乐，给人们以强烈的视觉冲击力。多样化的纹理和颜色，闪闪发亮的反射面是必须要有的。单独来看，它看起可能像一堆杂乱的拼贴画，但是作为一个整体，它却是构成一个美妙夜晚的基础。这些空间表达且仅表达一种感觉——一场全力以赴的盛会！

当你设计一个受中老年人（40到60岁）青睐的Flex俱乐部的时候，预期和经历就会有很大的不同。为了节省空间和节约项目成本，一个空间多种用途是非常有利的。许多空间白天作为餐厅使用，到了深夜11点以后则变成了俱乐部。作为餐厅，海景可以创造浪漫的就餐环境。晚间这些景色变暗，外部的窗户都关闭了，因为夜总会的目标就是创造喧闹中的"窥视"元素。最重要的是感受当下。Flex俱乐部白天时很难控制能量、灯光及氛围，而夜总会通常都是一个封装的没有窗户的空间，它的地理位置就不那么重要了。显然不用像一个高档餐厅的地理位置那样。许多俱乐部白天都是柔和浪漫的，随着夜晚的推进开始推向热烈的氛围。音乐或者娱乐节目也是由稍慢的开始，渐渐的推向热烈。这种俱乐部打烊也更早一些。夜总会有DJ，而Flex俱乐部则要有钢琴演奏或者现场乐队表演。可以进行互动娱乐，但却不像DJ俱乐部那样强劲。

夜总会中所使用的材料质地及颜色是灰暗的，可以有效的遮挡住顾客对陈设布置所造成的损坏，因为许多人会在展台或者桌子上跳舞，高跟鞋很容易将织物踩出洞或者将玻璃桌子打碎。需要考虑每一种材料的持久性和美观性。当你设计一个Flex俱乐部的时候，要注意当其作为餐厅使用的时候，材料需要看起来崭新而丰富，且要保证夜晚作为俱乐部使用的时候，材料足够结实。在餐厅营业期间，灯光要明亮且材质要以一定的方式呈现。一旦进入到夜晚，灯光要变暗，材质将失去其光泽。还需要融入稳重高雅的色调以与其他五星级就餐服务相媲美。所用材料必须要与俱乐部本身一样有视觉上的灵活性。需要做好维修及更新的预算，否则空间很快会显得陈旧。

Flex俱乐部就像变色龙一样一直在不断的变化中。在Flex俱乐部中，当夜总会元素出现的时候，所有的餐厅经营元素都将消失。餐桌被挪走，吧凳也被从吧台处挪走，换之以娱乐元素。还需要一个大型的存储区和为跳舞、灯光设备及大型音响系统准备的区域。所有区域必须要在白天的时候"消失"。这些元素在Flex俱乐部中是相对不太重要的，但是当顾客想要体验夜生活的美妙时，它又必须要完美的呈现出来。

这些俱乐部也会租作会展或者酒店特殊活动时使用。增加收入的机会是无限的。Flex俱乐部因其巨大的吸引力和灵活性，可以成为产业的名片。

希尔顿俱乐部是Flex俱乐部的一个典型例子，已经建立了很好的就餐声誉，且已经成为酒店内夜生活的基石。巧克力色、蓝色、绿色和米色在白天显得很高雅，在夜晚又显得幽暗。顾客能触摸到的地方所用材料都很舒适也很结实。墙的颜色往往比较淡一些，这样俱乐部的灯光可以真实的照射在上面。装饰灯具非常引人注目且规模巨大，能很好的填补巨大的空间。后面的淡玛瑙色吧台稍稍隐藏在吧凳的后面，但当夜间吧凳被挪走后又变得活力四射。木制地板使舞池的地点具有了灵活性。钢琴演奏区作为音乐娱乐中较轻缓的一面表现得恰到好处。当乐队或DJ演奏的时候，人们的焦点主要集中在舞池上。当作为夜总会功能使用的时候，就餐区被租赁成了贵宾区。贵宾们对酒水需求量最小，但带来的收益却是最大的。这些区域通常提前几个月就被租赁出去。经营者必须要灵活应对这种问题且在解决使用问题上要有创新性。当然一个企业的经营特性绝不会使Flex俱乐部处于闲置状态。

Hilton Player's Club
希尔顿俱乐部

Location:
Hilton Hotel and Resort, Atlantic City,
New Jersey, USA

Designer:
WESTAR Architects

Photographer:
Darius Kuzmickas

Area:
548 m²

Completion date:
2006

项目地点：
美国新泽西州, 亚特兰大市, 希尔顿酒店度假村

设计师：
WESTAR 建筑师事务所

摄影师：
达赖厄斯·库兹米卡斯

面积：
548平方米

完成时间：
2006

The challenge presented was to create an exclusive club that caters to the expectations and high standards of their discerning customers. Recognising that competition for repeat customers is great and enticing new customers to their property is critical, the results of this remodelling were that the Club was able to maintain its existing players as well as attract new customers.

The multiple facets of this project included creating nine areas, each with unique experiences but not so disparate as to cause unease but rather create a flowing transition: the entryway, host podium, queuing area, piano lounge, bar, casual lounge, two dining areas, and buffet station – all overlooking the Atlantic Ocean. Multiple visits by customers created the need for varied experiences throughout the space. Social or intimate, quiet or musical, focal seating or seating with a view – all options were addressed and each space was designed so that customers would be able to choose a seating area where they could unwind after a busy day.

The colour palette uses powder blue and green as accents against a background of chocolate and beige. The richness of the palette complements the preferred image of the customer's desires. The materials used were chosen for their richness and quality. Brazilian blue stone, backlit onyx bar panels, antiqued

设计此项目所遇到的挑战是如何创建一个可以为其高素质顾客提供高标准服务的、独一无二的俱乐部。当时意识到争取回头客的竞争很激烈且吸引新顾客很关键，因此这一改造的结果是俱乐部既可以留住现有的顾客，同时又可以吸引新顾客。

该项目的设计包含多个方面，其中包括建立九个带有独特氛围又并非毫无关联的区域，既要避免引起顾客不安的情绪，又要体现流畅的空间过渡：入口通道、接待处、等候区、钢琴休闲吧、酒吧、休闲厅、两个就餐区和一个自助餐厅——在所有这些区域中都可以俯瞰大西洋的美景。由于顾客到访甚多，因此很有必要在各个空间中营造不同的氛围。社交的或私密的，安静的或配有音乐的，集中就座或者分散就座的——各个空间都有不同的氛围供顾客选择，因此在劳累忙碌了一天之后，顾客可以选择一个可以放松身心的就座区。

采用蓝色和绿色作为主色调与巧克力色和米色的背景相呼应。色调的丰富性成为了顾客的首选。使用的材料也是根据其丰富性和质地而选择的。巴西蓝石、背光玛瑙板、仿古金箔柱、绒面墙、皮革、棕榈木地板、水晶球灯、15英寸高木及树脂同位孔灯、玛瑙灯光装置和独特的Player Card艺术品都具有最高品质且客户皆可触及。顾客群体会根据自己家中的装饰来品评俱乐部中所采用的材料。因此，俱乐部会吸引顾客在此停留、休息及放松。

1. Leather sofas increase elegant atmosphere of the lounge
2. Light of pendant lamps and light points creates a warm space
3. Round benches and tables show the high quality of the space
4. There are different seating areas in the restaurant
5. The piano lounge shows elegance
6. The flooring adds to the dignity of the space
7. The bar is shining with yellow light

1.皮质沙发为休闲吧增添了高雅的气氛
2.吊灯和点灯的光芒为空间增添了温馨感
3.环形桌椅显示了空间的高品质
4.餐厅中有不同的就座区域
5.钢琴休闲区充满了高雅的气息
6.地板设计为空间增添了高贵的气质
7.吧台闪耀着黄色的光芒

gold leaf columns, suede walls, leathers, palm tree wood floors, crystal ball lights, 15' tall wood and resin peek-a-boo lights fixtures, onyx light fixtures and unique Player Card Art are all of the finest quality and are within the touch of the clients. The age group of the customers responds to the choice of materials with a familiar knowledge, yet an unattainable level of finish in their own homes. The Club entices customers to stop in, relax, and unwind.

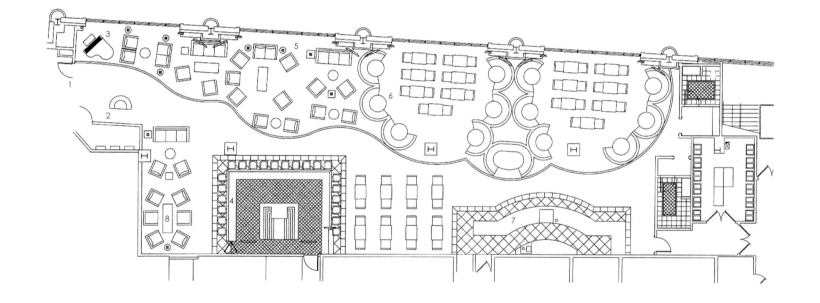

1. Entrance	1.入口
2. Host podium	2.接待处
3. Piano lounge	3.钢琴休闲吧
4. Bar	4.酒吧
5. Ocean view lounge	5.观景休闲吧
6. Booth	6.休闲厅
7. Display kitchen	7.展示厨房
8. Wood room	8.木室

6

7

Lounge Design
– By Paul Heretakis

The modern-day lounge is often used as a catch-all gathering space for the weary in order to find decompression and rejuvenation. Food and beverage offerings are simple and small as is the more laid-back entertainment programme. The space is about the sociability of its guests. It is meant to be friendly and interactive. Guests that want privacy will have to search out the few dark corners to get lost alone with their inner thoughts. The space is meant to function as a meeting place before and after any event. Many guests use it as their sociable home base with multiple visits each day. When not in their hotel room they can be found in the lounge.

The spaces must be luxurious, sophisticated and glamorous. Lots of WOW design features, reflexivity and sparkle. Rich colour schemes that balance colours and contrast in hues. It is used throughout the day for many purposes and must be memorable but not fully absorbed on the first visit. Each subsequent visit should uncover new details and features not seen before. Fresh floral presentations and a uniformed staff can alter the appearance of the room's presentation throughout the day. The colour schemes must be interesting yet subtle enough not to take the room over nor date the space quickly. Rich materials that are classic such as leather, marble and woods should be used generously. The light

fixtures must be custom and unique to the room. Lighting can date a space quicker than any other material, so the design must be timeless. Motifs of luck, fortune and long life are incorporated to enhance one's inner self throughout the visits.

The lounge is usually split into rooms within rooms. A bar area is a must for those that want to enjoy a drink and some conversation before moving onto the rest of their evening. The bar is often large in order to serve groups quickly and accommodate various clicks of people. The bar is also the final resting place for many of those who went out for the night but come for the last night cap drink before moving onto their hotel room. The lounge area is for the gathering of the social masses who want to see and be seen. They tend to stay for long periods of time and for various reasons. At certain times it is a communal meeting place to start or finish the day while other times it is the chosen destination. Of course the final area is the dining zone, where people use the space for its cuisine offerings. Great light-small plate meals are served in an atmosphere that is more casual and accommodating than a stuffy 5-star restaurant. True diners and multiple courses are replaced with buffet or table-presented servings of small-plate offerings. Light bites that intrigue and tantalise. A more common option to the heavy diner feeling most people avoid in today's more heath-

oriented approach to dining. A looser, more casual option to dining and a place where you can retreat to the bar or lounge after the meal in order to enjoy the entertainment. Continental breakfast can also be served during the morning hours when the rooms are lit with the glow of the morning sun. The sun will wash out some of the sophistication of the room and act as the rays of life in order to enhance the start of your day with a bright and cheery outlook. During the lunch hours light buffet offerings complement the beverage programme and create a gathering place for those looking for a place to decompress from the activities from the day.

The entrainment varies from Jazz quartets, piano players with a complementing singer and lively swing bands. The closer you are to the entertainers, the more they become part of your evening, while other guests will be in parts of the lounge that are not overcome with the sounds. A variety of experiences within one lounge is always the goal. During the morning and daytime hours computers are the preferred instruments of entertainment, but once night falls, all interests turn to a let-loose attitude of fun first.

The Bally's Lounge is influenced by the Americana style with its masculine oversized furniture. The colour palette goes from rich reds and coppers in

休闲吧设计
——保罗·哈里塔基斯

the bar to the brighter and friendlier blues and golds in the dining area that faces the windows. Four double-sided fireplaces create eight seating areas washed in the warmth of the traditional hearth. They also separate the lively bar from the more subdued lounge area. The feelings of comfort are meant to be felt immediately upon entry into the room.

The Caesars Lounge is more European and sophisticated. The furniture is more formal and feminine. The colours are brighter and the materials are more expensive and impressive. The room is purely 5-star as is the hotel itself. Dining area enjoys the view while the lounge areas are broken into little jewel box settings of crotch-cut book matched wood and marble columns. Colonnades and gold leaf impress form the moment you enter the space. It is a tour de force of design that is meant to uplift your style.

现代休闲吧通常被作为疲惫的人们聚集、寻求放松、恢复活力的地方。食物和饮品供应相对少且简单，悠闲的娱乐节目相对较多。这样的空间是客人们的社交场所，气氛要柔和且要有互动。需要独处的客人不得不寻找那些暗角处来进行内心的思考。休闲空间的功能是作为活动前后人们相聚的地方。许多客人只把这里当作是社交基地，每天在这里会见许多人。如果他们不在酒店房间内，就在休闲吧里。

休闲空间一定要豪华、高雅、迷人。要有许多令人叹服的设计特色，要闪耀着光芒。丰富的色彩设计既要保持色调平衡又要形成鲜明对比。颜色处理既要让人记忆深刻又要让人第一次看到后不能完全领会其中的妙处。要让客人在后续的每一次来访中都感受到新意。鲜花装饰和整齐穿戴的工作人员可以改变室内空间给人的印象。色彩搭配一定要足够精妙有趣，既不要使空间太过耀眼也不要使空间很快过时。一定要慷慨的使用各种经典而丰富的材料，如皮革、大理石和木材。灯光设置一定要独特且与室内设计相配合。灯光比其它材料更容易使空间过时，因此，设计一定要经久不衰。好运、财富和长寿的主题综合起来让来访的客人们感受强烈。

休闲吧通常设计成室中室的形式。酒吧区是必须要为那些想先享受一下饮品、进行一番交谈的人们准备的。酒吧通常很大，这样可以高效率地为顾客服务，满足顾客的需求。对于那些夜晚从外面回来的人，在回酒店房间之前，酒吧则成了他们想要享受小酌的最后一站。休闲区也成了那些想要看到别人也希望被别人看到的人们的社交场所。他们往往因各种原因在这里停留很长一段时间。有时候这里只是人们在一天的开始或者结束时进行聚会的场所，但有时候，这里则是人们理想中的目的地。当然，最终的地点是就餐区，在那儿人们可以享受美食。明亮的灯光、精致的餐点以及就餐的氛围都要比沉闷的五星饭店更加轻松而惬意。真正的大餐和各式菜肴都替换以自助餐或者小盘菜肴。便餐很吊人胃口。在主张健康就餐的今天，人们都避免大吃大喝，这是大多数人的就餐选择。这里的就餐环境轻松随意，饭后还可以去酒吧或者休闲吧享受娱乐节目。当早上

第一缕阳光照亮室内的时候，这里也会提供欧式早餐。阳光洗涤了室内的烦杂，就像生命之光一样用明快和欢乐照亮了人们美好一天的开始。午餐时段有自助餐和饮料供应，为上午繁忙活动归来寻求放松的人们提供了很好的聚会场所。

娱乐活动从爵士乐四重奏、歌手演唱钢琴伴奏到生动的摇摆乐队表演应有尽有。你离表演者越近，娱乐就会成为你夜晚中越重要的组成部分。而其他客人更多的会在休闲吧里度过，这里是音乐及一切声音触及不到的地方。人们的目标总是希望在一个休闲吧里可以有各种各样的体验。早上和白天的时间里电脑是人们娱乐的主要工具，但是当夜晚降临的时候，所有人就都想先轻松娱乐一番。

巴利俱乐部受到了美洲那种伟岸大气风格的影响。色调从酒吧里丰富的红色和铜色到面对窗户处的就餐区中更加明亮而柔和的蓝色及金色。四个双面壁炉形成了8个可以享受传统炉火温暖的就座区。人们一进屋就立即能感受到舒适与温暖。

凯撒俱乐部设计则更加具有欧式风格且更高雅精致。家具陈设更加正式而柔和。颜色更加鲜亮，材料更加昂贵、令人难忘。俱乐部像酒店本身一样具有五星级品质。就餐同时可以享受美丽的景色，休闲区被分成了小型的珠宝盒式的空间，配有整齐排列的木制和大理石制成的圆柱。一进入室内，石柱廊和金叶子图案会给人们留下深刻印象。来到这里感受设计的力量，可以提升您的品位。

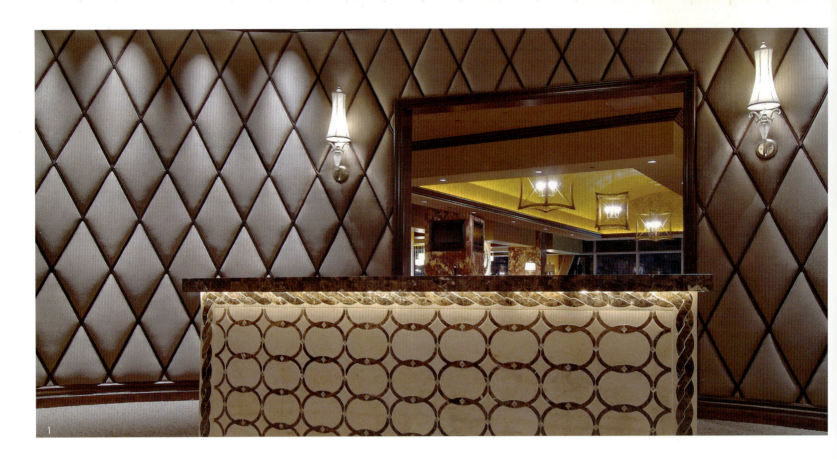

Bally's Player's Lounge

巴利俱乐部

Location:
Atlantic City, New Jersey, USA

Designer:
WESTAR Architects

Photographer:
Darius Kuzmickas

Area:
1,765 m²

Completion date:
2007

项目地点：
美国，新泽西州，亚特兰大市

设计师：
WESTAR建筑师事务所

摄影师：
达赖尔斯·库兹米卡斯

面积：
1765平方米

完成时间：
2007

The design objective was to create a style within the Americana genre that would establish the Bally's brand. Upholstered walls and views of the Atlantic Ocean overwhelm customers when they enter the Club.

The use of columns, fireplaces, low walls and niches was required to subdivide the 1,765-square-metre space and 425 seats into 18 unique, intimate seating vignettes of less than 25 seats each. A 45-seat upholstered tufted leather bar, complete with crystal buttons, overlooks a historical park, while the lounge area is bordered by four double-sided fireplaces.

Overstuffed furniture, living room settings, wall shelving and unique Americana artifacts give guests a sense of home. The liberal use of gold, silver and copper leaf graces the ceiling; rich burgundies, browns, reds and chocolates make up the colour palette. The ceiling design uses various applications of copper ceiling tiles, wood beams; embossed wall covering, mirrored panels and a gold leaf vault add dimension to the ceilings. Hand-applied gold-and-copper leaf over a background of burgundy adds a warm finish to the walls. Faux suede covers other walls to create a soft, intimate environment.

The most sought-after seats are along the 60m expanse of floor-to-ceiling windows that

设计目标是要在美洲流派内创造一种可以打造巴利品牌的风格。当客人进入俱乐部时会被豪华的墙面和大西洋的美景所倾倒。

需要使用圆柱、壁炉、矮墙和壁龛来将这1765平方米、容纳425个座位的空间分隔成18个独特的、私密的、少于25个座位的小空间。一个能容纳45个座位的皮革软垫装饰的酒吧，配有水晶装饰，可以俯瞰历史公园。休闲区域周围配有四个双面壁炉。

加有厚软垫的家具、客厅布置、墙面搁架和独特的美洲文物都给客人以家的感觉。金银铜叶子装饰不拘一格，给天花板增添了高雅的气质；丰富的红葡萄酒色、棕色、红色和巧克力色构成了主色调。天花板的设计采用了各种铜制的天花板材料和木梁；浮雕壁布、镜像板和金边拱顶增加了天花板的维度。手工制作的金和铜叶子装饰覆在红葡萄酒色的背景上，给墙面增添了温馨感。人造鹿皮覆盖着其他墙面，形成了柔软私密的环境。

座位后面最抢眼的地方是沿着地板延伸61米一直到天窗的那部分，可以俯瞰大西洋。传统美式风格的镶嵌式地板点缀着四个不同类型的木制地板和三个不同类型的石材地板。每种材料的运用都为顾客创造了一个独立而独特的就座空间。美洲风格的奢侈品给客人以舒适的享受且与他们的兴趣相一致。

1. The magnificent sight of the reception
2. High-quality bar and benches
3. Exquisite furniture and decorations create an exclusive space
4. The bar and sofas create a harmonious atmospheres
5. Different seating areas with different atmospheres
6. Private lounge with exquisite design
7. Elegant colour increases dignity for the space

1.华贵的接待处
2.高品质的吧台和吧凳
3.精致的陈设装饰点缀了独特的空间
4.吧台与沙发为室内增添了和谐的气氛
5.不同的就座区有不同的氛围
6.私人休闲室设计精美
7.典雅的色调增添了室内的高贵气质

overlook the Atlantic Ocean. A mosaic floor depicting traditional American patterns is interspersed with four different types of wood floors and three different patterns of stone floors. Each application of material creates a separate, unique seating experience. Americana luxury was a comforting offering to the guests and in line with their interests.

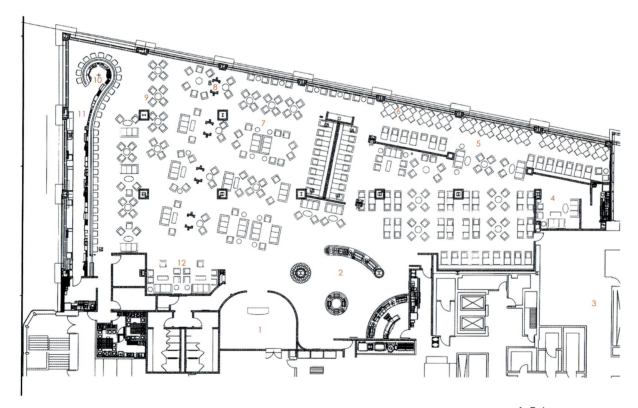

1. Entrance	1.入口
2. Buffet service	2.自助餐服务台
3. Blue dining area	3.蓝色餐厅
4. Private lounge	4.私人休闲吧
5. Copper dining room	5.铜色餐厅
6. Ocean-front view	6.观景区
7. Gold dining room	7.金色餐厅
8. Fireplace seating	8.壁炉区
9. Lounge seating	9.休闲就座区
10. Dome above	10.圆顶
11. Bar	11.酒吧
12. Library	12.图书室

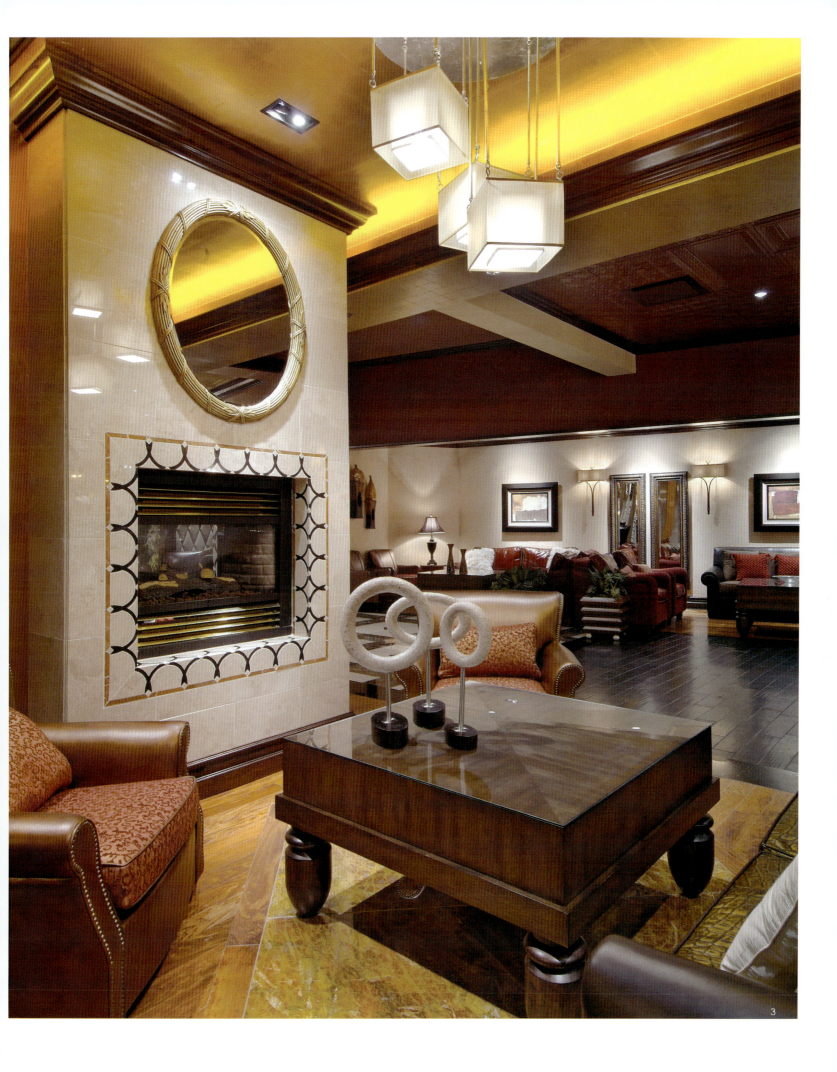

6

7

Caesars Player's Lounge
凯撒俱乐部

Location:
Atlantic City, New Jersey, USA

Designer:
WESTAR Architects

Photographer:
Darius Kuzmickas

Area:
1,300 m²

Completion date:
2007

项目地点：
美国，新泽西州，亚特兰大市

设计师：
WESTAR建筑师事务所

摄影师：
达赖厄斯·库兹米卡斯

面积：
1300平方米

完成时间：
2007

Caesars Player's Lounge was designed to replicate an elegant Roman Villa, sparing no expense. As Caesar's moves forward into the 21st century with enhancements to their property, their objective was to establish their brand identity and its association with elegance, opulence, and affluence. Guests are quickly transformed into another world of grandeur and timeless exclusivity.

Byzantine movement sets the tone as guests follow pathways that are bordered by a marble colonnade that creates focal vistas. The designer travelled to Spain and Italy to personally select the stone columns used in the colonnade. Over 270 square metres of hand-laid gold leaf graces the barrel vault entry and pathway's ceilings, while book-matched golden onyx creates a backdrop for the host.

The rarest of pink onyx is backlit behind the bar and creates a visual end to the entry procession. The 1,300 square metres of marble floors use six different patterns, four different mosaic medallions, and over twenty colours of stone. The ceiling is framed with mahogany crown molding stained and painted with gold accents. Authentic Roman ceilings were meticulously recreated. Gold pearlised Venetian plaster, book-matched red travertine, and crotch-cut mahogany walls create an overwhelming feeling of rarity and richness.

凯撒俱乐部的设计是为了不遗余力的复现一座优雅的罗马别墅。随着凯撒产业的增加，凯撒酒店向前迈进了21世纪，其目标是树立自己高雅、华贵、丰富的品牌形象。来到这里的客人们会迅速进入到另外一个独一无二、富丽堂皇的世界。

拜占庭风格为俱乐部定下了基调。过道两边是大理石柱廊，为客人创造了华美的画面。设计师亲赴西班牙和意大利挑选石柱来装饰柱廊。270多平方米手工制作的金叶子图案装饰为入口拱顶和过道处的天花板增添了高雅的气息，而呈书页式排列的的金色玛瑙材料则为主人创造了华美的背景装饰。

罕见的粉红玛瑙材料从酒吧的后面照射出光芒，让客人无法看到酒吧深处。1300平方米的大理石地板采用了六种不同的形式、四种不同的图样和20多种石头的颜色。天花板四周是红木框，配有金色花纹。复现了真正的罗马天花板。金色泛着珠光的威尼斯墙粉、书页式拼接的红色石灰华和整齐的红木墙让人产生一种珍贵而丰富的感觉。

家具陈设的设计灵感来自意大利时期的创作作品，为空间增添了温馨和真实的色彩。装饰品由意大利最好的品牌如范思哲、阿玛尼所提供。植物装饰为红、蓝、金色，增加了色调的丰富性。雕刻和蚀刻玻璃增添了空间的现代主义色彩。

1. Grand reception
2. The magnificent entrance welcomes guests with dignity
3. The design of the lounge provides guests with comfort and warmth
4. Sofas and ceiling create a comfort space
5. Fireplace seating area provides a warm atmosphere
6. Soft sofas and warm settings match with each other
7. Buffet with fancy dishware

1.华丽的接待台
2.入口处的华贵欢迎着客人的到来
3.休闲吧的设计为顾客提供了舒适与温暖
4.沙发和天花板交相辉映
5.壁炉就座区为顾客提供了温暖的氛围
6.柔软的沙发和温暖的背景相互呼应
7.自助餐厅摆放着精美的餐具

The furniture is inspired by Italian period pieces and adds warmth and authenticity to the space. The accessories were provided by the finest Italian boutiques, from Versace to Armani. Florals add a burst of colour to the richness of the red, blue, gold colour palette. Carved and etched glass lends a complementary touch of modernism to the space.

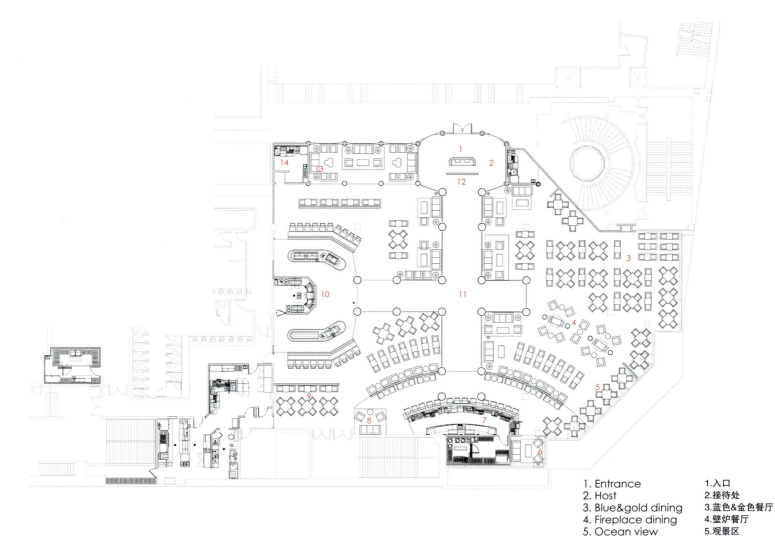

1. Entrance	1.入口
2. Host	2.接待处
3. Blue&gold dining	3.蓝色&金色餐厅
4. Fireplace dining	4.壁炉餐厅
5. Ocean view	5.观景区
6. President's room	6.主席办公室
7. Bar	7.酒吧
8. Lounge	8.休闲吧
9. Red&brown dining	9.红&棕餐厅
10. Buffet	10.自助餐服务台
11. Grand walkway	11.大走廊
12. Onyx wall	12.缟玛瑙墙
13. Caesar's room	13.凯撒休闲室
14. Service station	14.服务站

6

7

Creating the Privium Experience
– By Drs. Conny Lanza, General Manager of Amsterdam Schiphol Airport

Enlightening experiences: A good brand reaches both instrumental and emotional values and shows itself to be consistent in all its different manifestations: the product, the service rendered, the corresponding room, the building.

It is important that everything tallies. In short, the different shapes in which a brand manifests itself, need to be consistent and need to reinforce each other. The introduction of the Privium ClubLounge, designed by M+R, is a milestone in the development of Privium, the service programme with the revolutionary iris scan, which offers a range of privileges for frequent fliers at Schiphol Airport. The lounge touches the deeper needs of customers and is an interpretation of the Privium brand values.

Products and Services must be authentic and sincere and in line with the brand values. Only then will the customer actually appreciate them. If a brand does not respond to any deeper needs and if the product, the service or relevant area or building does not touch any deeper values, then everything remains very superficial.

An unusual dream: In the briefing that accompanied the assignment for the design of the ClubLounge, not only the background, aims of the project and a description of the functional

28 - 29

and qualitative requirements were included, but also a lot of attention was paid to the emotional state of frequent fliers, their personal characteristics and the Privium brand values. M+R was explicitly asked to offer added value, to add an additional dimension to the design.

Travelling by plane is a form of expansion: a broadening of your experiences, broadening your horizon. Within the travel process, the airport is the springboard to another world, culture, climate, time zone. There is no final destination, only a halfway station, a "pit stop", a "nowhere town". In the course of their travel process the frequent fliers experiences tension between positive feelings of excitement, of exploring the world and negative feeling of loss of control, insecurity, tiredness and restriction of their freedom. Frequent fliers are people who are used to being in control, taking control.

The central theme is that in the "theatre" of the Privium ClubLounge the Privium members play the leading part. They are the centre of attention. They are "on stage". Then there were also the following points listed in the briefing:

• Because of the stressed state travellers are in when they travel, it is a requirement that the Privium ClubLounge be spatial, light and

uncluttered. Only then can frequent fliers relax and find their equilibrium.

• Privium members are part of a select group. Privium offers a visible distinction. The Privium ClubLounge offers privileges that are only accessible for Privium members. They can recognise themselves in the Privium ClubLounge, which will make them feel special and they will be given the recognition they deserve.

• Travelling is often a lonely phenomenon and international travel is often accompanied by a faint feeling of being uprooted. Privium offers an international platform where different nationalities, residents of different countries and different cultures can meet. In the ClubLounge a frequent fliers feels connected, and is prepared to be more open to others.

• The Privium ClubLounge must have sufficient stimuli to compensate for the missing display of platforms, views and light. So it must be special, spectacular, and varied in shape and light. This will allow a frequent fliers to feel more impulses and thus "alive" and "part of the global world".

The requirements for and expectations of the Privium ClubLounge design commission were high. M+R won the pitch from five other renowned agencies, but it not just met the expectations; it exceeded them. Not

创造Privium式体验
——康尼·兰扎（斯希波尔机场Privium俱乐部总经理）

启蒙体验：一个好的品牌不仅具有工具价值，而且具有情感价值，且在其不同的表现形式中展示了各个方面的一致性：产品、服务、室内、建筑。

一切都与之相吻合是非常重要的。简而言之，品牌的不同表现形式要一致统一，相辅相成。由M+R设计公司设计的Privium休闲俱乐部是Privium品牌发展过程中的里程碑，是利用虹膜扫描为频繁往返于斯希波尔机场的旅客们提供的特别待遇的服务项目。俱乐部满足了顾客内心的需求，诠释了Privium的品牌价值。

产品和服务一定要真实诚恳，且与品牌价值相一致。只有做到这一点，顾客们才会真正的满意。如果一个品牌没有与顾客的内心需求保持一致，如果产品、服务或者相关的空间或者建筑没有触及到任何深层价值，那么一切都会显得很虚伪。

一个非同寻常的梦想：在休闲俱乐部的设计简介和任务要求之中，不仅包含了设计背景、项目目标以及对功能和质量的要求，还包括了对顾客情感上的考虑、他们的个性特征以及Privium的品牌价值。M+R公司还被明确要求提供附加价值——在设计中形成另外一个维度。

乘坐飞机旅行是一种扩张形式：增加你的阅历、拓宽你的视野。在旅行的过程中，飞机是带你通向另一个世界、文化、气候及时区的跳板。没有终点，只有一个中转站，一个"紧急加油站"，一个"任何地方都不是的城镇"。在其旅行的过程中，频繁旅行的旅客经历了探索世界那种积极兴奋的感觉，也经历了失控、不安全感、疲惫和压力等消极的感觉。经常乘坐飞机的人们习惯了控制与被控制。

设计的中心主题是要让Privium的会员们在这个"剧场"中起主导作用。他们才是视线的焦点。他们位于"舞台之上"。在项目要求中还有以下几点：

由于旅客在旅行过程中处于紧张状态，因此休闲俱乐部必须要宽敞、明亮、整洁。只有这样才能使旅客得到放松，心情平静。

Privium的会员们都是精心挑选出来的。Privium为其提供了与众不同的服务——只有会员们可以通过的特殊通道。这样使他们意识到自己身处Privium俱乐部中，使他们感觉自己很特别，这也是他们应该得到的。

旅行往往是孤单的，国际旅客通常会感到失去了归属感。Privium为顾客提供了国际性平台，不同国籍、居住于不同国家、拥有不同文化的人们在这里相遇。在这里，旅行常客会感觉到归属感，与其他人相处得更加自然。

Privium休闲俱乐部必须要有足够的刺激元素来补偿舞台、景观和灯光的空白。所以这里一定要特别而且壮观，设计形式和灯光要有所变化。这会使旅客感觉更加生动刺激，成为"全世界的一部分"。

对Privium俱乐部设计的要求和期望非常高。M+R在与其他五家知名机构的竞争中脱颖而出。它的设计不仅达到了设计的预期效果，还远远超越了期望值。M+R公司不仅满足了设计的众多要求，实际上还为设计增加了额外的维度，更加突出强调了设计理念。

平静、接触、改变：作为Privium的会员，人们有一种优越感。这种感觉通过顶级的陈设设计得到加强。著名的蛋形椅、天鹅椅、泡泡椅、索利泰尔和科布西埃躺椅等，每个设计都源于20世纪20年代、50年代和60年代的设计经典。即使是繁忙时段，休闲吧内也很安静。M+R设计公司通过使用厚毯和浅色遮光窗帘使室内空间整洁优雅。会员们将其称作"白色休闲吧"。这种设计给人一种宁静的感觉。休闲吧邀请您来这里放松休息。

相连工作区和就座区的结构使您在想休息的时候随时可以进入。同时为孤独的旅客提供与其他人见面的机会。从任何一个就座区都可以将其他区域一览无余。从任何一个位置都可以看到其

only did M+R meet the broad range of needs and requirements, they actually added an extra dimension, thus enhancing and reinforcing the concept.

Peace, contact and a change: As a Privium member, this gives you the feeling you are special. This feeling of exclusivity is emphasised by using top design furniture, with designs of the famous Egg Chair, Swan Chair, Bubble Chair, Solitaire and Le Corbusier's chaise longue, each one a design classic from the twenties, fifties and sixties. Even when it is busy the lounge emanates tranquility. M+R achieved this by creating space and orderliness and by using thick carpets and pale shades. Already members call it the "white lounge". This gives it an almost serene atmosphere. The lounge invites you to take it easy.

The configuration of the linked working and seating areas allows you to withdraw should you want to. At the same time it offers even the most solitary traveller the opportunity to meet others. From any sitting area you can oversee the other areas. From any position you have a view of other Privium members, or your co-travellers. This gives a sense of contact with other travellers, without the need to interact.

It makes the act of travelling and in particular your stay in the lounge less lonely, so travellers will not feel terribly uprooted.

Even the fixed Internet areas have been deliberately positioned so that they are not facing away from other Club members. And naturally, as a metaphor for the "connecting statement", there is the long centrally positioned bar. This is a location that is especially ideal for making contact with or feeling you are connected to others also sitting at the bar. Variety is achieved through the different sitting areas, each with its own function, atmosphere and design. There is an area with a reading table, a thick-carpeted relaxing sitting area with a fire, the "hanging bubble chair" section and the fixed Internet places. The extra dimension contributed by M+R is the organic character of the lounge. Nothing, absolutely nothing in this room is straight. Outdoor elements have been brought inside, such as the enormous bamboo stalks that seem even longer because of the mirror positioned above them, the fire in the fireplace and a stone wall that simulates outside light shining in. All this makes up the water, fire, air and earth elements that are an important part of the design.

Authenticity and consistency: Other decisive elements for the lounge experience are the smell, the food & beverages concept, the lighting, the staff and the service. A subtle aroma was chosen for the lounge. In the Light Energy Cabin there is a wellness scent. Even the Food & Beverage concept links up with the brand values. For example, the ever-popular Chardonnay wine is not served, only more special wines such as a Gruner Veltliner and a Sotoverde, which are greatly appreciated by connoisseurs. There are authentic French specialities such as Lenotre croissants and macarons.

A daring transparent light blue, almost silver, shiny colour was picked for the uniform, which matches the concept.

In short, everything is right. This is the power of the Privium ClubLounge. With gratitude to M+R.

他成员或者您的同伴。这种设计使人们不用进行互动就能感觉到与其他游客有着某种联系。使旅行与在休闲吧内的停留变得不那么孤独，这样一来，旅客们将不会感觉十分的失落。

固定的网络区也经过了精心的设计，俱乐部会员随时可以上网。长长的中央吧台自然就成了人们联系的的象征。这种位置安排非常适合会员们之间相互交流，使同坐在吧台附近的旅客们有种互相联系的感觉。不同的就座区域使室内变得多样化，每个就座区都有其特有的功能、气氛和设计形式：配有书桌的就座区；铺着厚厚的地毯，配有壁炉的区域；"悬挂式泡泡椅"区域和固定的上网区等。休闲吧内的多维度效果是M+R公司精心设计的特点之一。这个空间内绝对没有任何东西是笔直的。室外元素被引入室内，如巨大的竹竿在上部镜子的反射下显得更长。壁炉里的火苗和石墙都是模仿从外面照射进来的光设计而成的。所有的这些构成了水、火、空气和土壤元素，是设计的一个重要组成部分。

纯正性和连贯性：其他决定休闲吧内氛围的元素是气味、食物及饮品、灯光、工作人员和服务。设计师为休闲吧选择了一种优雅的芳香气味。在光能室里有一股健康的芳香。食物与饮品也与品牌价值相联系。这里并不提供曾经非常流行的夏敦埃酒，只有更加特别的酒如格联纳威特林那葡萄酒Sotoverde酒，这是备受行家们喜爱的酒。这里有真正的法国特产如勒诺特尔的羊角面包和马卡龙甜点。

制服的颜色是大胆的近乎透明的浅蓝色——接近于银色的一种鲜亮颜色，与设计理念非常匹配。

简而言之，一切都是最恰当的。这就是Privium休闲俱乐部的魅力所在。感谢M+R。

1. Information desk for Privium Members entrance of the ClubLounge
2. The Entrance is like an "eye"
3. Exquisite design of the buffet and bar
4. The neat reading area
5. Lounge seating area with special hanging chairs and computer chairs
6. Comfortable lounge seating area
7. Comfortable chairs, fireplace, and natural elements such as bamboo and stone tiles in the lounge

1.俱乐部入口处服务台
2."瞳孔"形入口
3.自助餐台与吧台的精美设计
4.整洁的阅读室
5.就座区内独特的挂椅和电脑椅
6.舒适的休闲吧座椅
7.舒适的椅子、壁炉以及竹子和石砖等自然元
　素点缀室内

M+R interior architects is an international operating office founded by Hans Marechal and Marie-Louise Rooijmans. Their fields of activity often involve complex assignments such as converting and designing offices, airports, libraries, restaurants, hotels, theatres and shops. Among their design skills and core activities for building and interior architecture they are also involved with revitalising existing buildings and monuments in particular. The architects from M+R determine the form and content of each design assignment on the basis of the programme requirements. Creativity, functionality, sustainability and ergonomics are translated in a well-thought-out manner into a unique final product with an identity of its own. The power of a strong design is vision, innovation and the quality of realisation.

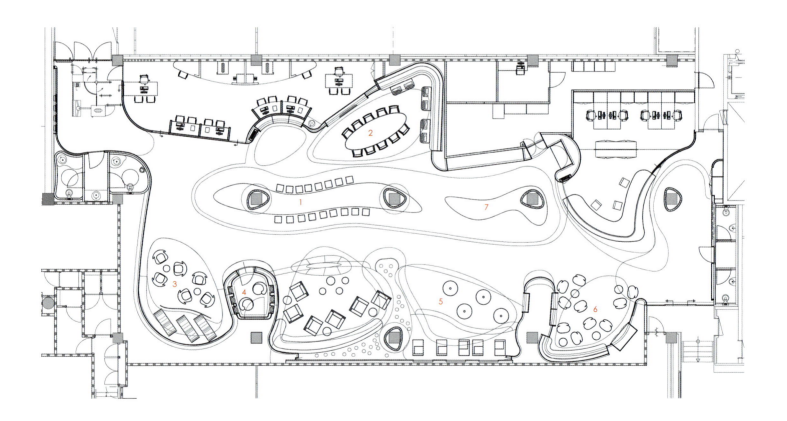

1. Buffet	1.餐饮服务台
2. Reading area	2.阅读区
3. Relaxing area	3.放松区
4. Light Energy Cabin	4.光能室
5. Lounge area	5.休闲区
6. Waiting area	6.等候区
7. Bar	7.酒吧

6

7

Responsiveness between Fashion, Art and New Culture
– By Joey Ho

We are living in an era that our relationships with our family and social structure are being challenged, and our role in the world is being reconsidered.

The impact of globalisation and technology changes the economical models. Bars and clubs are re-styled and contrived to a brand new environment. A new place, a destination where people can fulfill their emotional needs.

In this age of uncertainty, bars and clubs are spaces to help people recover their senses. These spaces enable our emotions to uplift our lives, liberate people's feelings and reveal aspects of their personalities that they don't know yet. Exceeding their basic function of entertainment & refreshment, bars and clubs are now transformed. They are places for parties and socialising that are

recognised in aspects of fashion and lifestyle. Designers for bars and clubs are now in fact looking for a new sensory design to provide emotional drivers. There are a few key drivers within us, other than our normal side of life:

Our drive for hope & engagement,
Our drive to escape,
Our drive to achieve glamour.

What lead us to a new dimension in emotional design in bars & clubs is "DESIGN". It is the ability to stir up emotion and help us to relieve forgotten sensations.

Today we define the meaning of our life in psycho-physical terms: what makes me feel good? That is the central question in our contemporary search for the meaning of our life. In Kulissen des Glucks' latest publication "Staging Happiness", he analysed

various kinds of urban or semi-urban settings created to produce these sensations: "They are consciously constructed façades, which are filled with life by suppliers and consumers." Bars & clubs are designated to generate a beautiful experience, a moment of happiness.

If the design created is unexpected by the user or adapted to uncommon categories or cross-boundaries, it can bring us new sensory discoveries for the foundation design of bars & clubs; it's not just a question of style, but a cultural shift. While it is true that the basic principle of bar & club design may be divided between fashion, art and new culture, in all these three areas, responsiveness is the key factor. In many ways, the designer's skill as an artist is to draw on a diversity of sources and experiences to achieve a successful result.

时尚、艺术与新文化的共鸣
——何宗宪

We treat bars and clubs as theatrical spaces like a stage that encourage performance. They are not a place simply for drinking and dancing any more, but a place for pleasure & rediscoveries of oneself. As a result, designers for bars & clubs have to experiment and push the boundaries in terms of themes and aesthetics.

我们生活在一个与家庭和社会之间的关系受到挑战的时代，我们在世界中所扮演的角色正在被重新审视。

全球化和技术的影响改变了经济模式。酒吧和俱乐部也改变了风格，营造崭新的环境。一个可以满足人们精神需求的新去处。

在这个变化不定的时代，酒吧和俱乐部是帮助人们恢复其感观的场所。这些场所使我们精神振奋地去生活、解放人们的思想、展现人们自己都没有意识到的个性。现今的酒吧和俱乐部超越了其基本娱乐身心的功能。变成了举办聚会和社交活动的场所，是在时尚和生活方式层面得到认可的流行聚集地。酒吧和俱乐部的设计师们实际上正在追求的是一种感观设计，为人们提供情感驱动力。在我们的生活中，除了普通的生活层面上，存在着关键性的驱动力：

我们心存希望，渴望参与，
我们渴望逃避，
我们渴望实现我们的魅力。

引导我们通向新的酒吧俱乐部情感设计维度的是"设计"。这是一种激发我们情感的动力，有助于我们恢复已被遗忘的感观。

如今，我们根据心理–物理双层面来定义我们

的生活方式：什么能使我们感觉舒适？这是我们在追求生活意义的过程中面临的核心问题。Kulissen des Glucks在其最新出版的作品《分期幸福》中，分析了多种可以形成这种感觉的城市或半城市化背景："它们的外表精心打造，里面充满了具有活力的供应商和消费者。"酒吧和俱乐部是明确带出美好体验和幸福时刻的场所。

如果设计出人意料、类别罕见或者跨界，那么它便可以为我们提供新的感观探索，来实践酒吧和俱乐部的设计基础；这不仅仅是一个关于风格的问题，而是一次文化转变。酒吧和俱乐部设计的基本原则可以在时尚、艺术和新文化之间划分，在这三个领域内，共鸣是中心要素。在许多方面，作为艺术家的设计师们都善于在多种资源和经历中描绘成功的蓝图。

我们把酒吧和俱乐部看成是戏剧化的空间，就像激励表演的舞台一样。它们不再是仅用来饮酒跳舞的场所，而是一个用于娱乐和重新寻找自我的地方。因此，酒吧和俱乐部的设计师们不得不试着打破主题性和美学之间的界限。

1

China Qiandaohu Country Club
中国千岛湖乡村俱乐部

Location:
Hangzhou, China

Designer:
Joey Ho / Joey Ho Design Ltd.

Photographer:
Chai Zhicheng, Bao Shiwang

Area:
4,334 m²

Completion date:
2007

项目地点：
中国，杭州

设计师：
何宗宪 / 何宗宪设计有限公司

摄影师：
柴之澄、鲍世望

面积：
4334平方米

完成时间：
2007

Enjoying the mystical beauty of the Thousand Islands Lake in Hangzhou, China, the interior design of this golf country club hinges upon the concept of golf greens. The designer translated this into a series of floating island space, which are juxtaposed upon one another to create an intriguing layering effect. Squarish patterns also abound throughout the space, and they are interpreted both two dimensionally (as wall panelling, ceiling plane, mirrored mural) and three dimensionally (as floating islands, benches and sofas). All these make up a spatial composition that stresses graphical layering, a notion that the designer has explored consistently throughout the project.

Water also emerges as a core concept inside the country club. Set against a beautiful lakeside backdrop, the architecture is surrounded by the tranquility of water. By introducing water feature into the interior, the designer set out to blur the boundary between inside and outside. In the reception hall, water is expressed as pockets of ponds, a form of negative space to link up the floating islands. The interplay between solid and void orchestrates an interesting pathway as one proceeds, and the experience of wandering amidst manmade ponds and natural lake is heightened to prepare one for the upcoming luxurious enjoyment.

此高尔夫乡村俱乐部的室内设计基于绿色高尔夫的概念，尽享了中国杭州千岛湖上的神秘之美。设计师将其划分成一系列浮岛空间，它们之间紧密相连，创造出一种耐人寻味的层次感。方形图案及装饰贯穿整个空间，无论从平面（墙板、棚面、镜像壁画）还是立体方面（浮岛、长凳和沙发）都得到了很好的诠释。所有这些构成了一个宽敞的空间，增加了平面的层次感，这是设计师在整个设计过程中不断探索出来的概念。

水也是乡村俱乐部项目概念中的一个核心元素。坐落于美丽的湖边，俱乐部建筑被水的宁静所环绕。通过在室内引入水景，设计师有意将室内和室外的环境相融合。接待大厅里的水元素被设计成若干小池塘，这是连接浮岛空间的一种形式。虚实之间的相互作用形成了趣味横生的过道，给漫步在人造池塘与自然湖景之间的人们提供了美妙的体验，为接下来的奢华享受做了铺垫。

1. The light colours create a neat space
2. The exclusive design forms are very impressive
3. Proper flooring colours match with the high ceiling
4. The ceiling with small lights is like sky full of stars
5. Interior light matches with the view of outside
6. Design of different foyers provides a continuous elegance
7. The overall design of the space gives a bright view

1.浅色调设计形成了整洁的空间
2.独特的设计形式非常令人难忘
3.地板的颜色与天花板完美结合
4.天花板上闪着小灯犹如繁星点点的夜空
5.室内灯光与室外景色交相辉映
6.不同层面的设计连续而典雅
7.整体空间设计给人一种眼前一亮的感觉

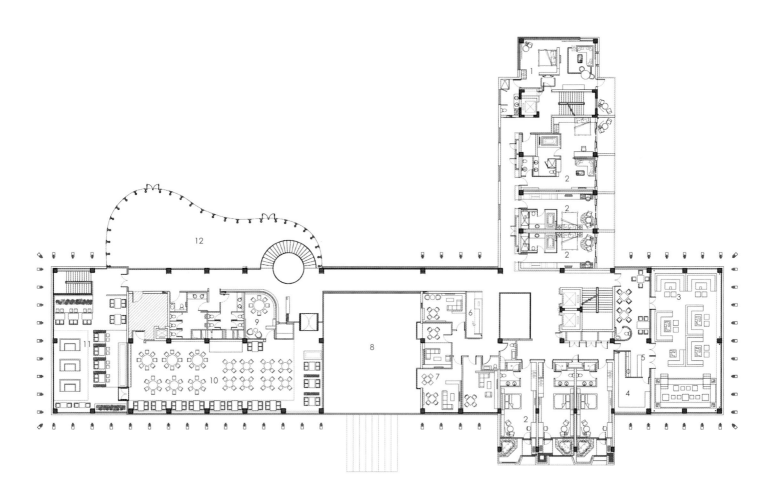

1. VIP room | 1.贵宾室
2. Guest rooms | 2.客房
3. Outdoor café seating | 3.室外咖啡茶座
4. Work station | 4.工作间
5. Bar area | 5.酒吧区
6. Service | 6.服务台
7. Card rooms | 7.棋牌房
8. Lobby | 8.门厅
9. Compartment | 9.包间
10. Chinese food dining | 10.中餐厅
11. Outdoor dining | 11.室外餐厅
12. Western dining&café | 12.西餐咖啡厅

Everlasting Space Design
– By Oobiq Architects

The designer's brief introduction of the DROP Shanghai Project:
The client asked us to create an interior that could be represented by a definition: contemporary classic. An interior that should have been trendy but not too fashionable. Too fashionable spaces are getting old soon, and our client wanted to create a club that could keep its images for a long time. So, starting from the concept of contemporary classic, we created all the interior design.

Detailed description of the project:
A "contemporary classic" design led us to the first choice in terms of furniture: Chesterfield sofa was the style for all the seating inside the Club. It is a classic design, but with a contemporary shiny colour, with a feeling that reminds people of Europe. An other important choice was related to the walls. Wooden boiserie is very traditional, but is detached from the wall, as floating elements. Another reminder of Europe. Then the floor, with a decoration that also became the idea for the packaging and graphic of the club. Approaching the floor we were inspired by the concept of the kaleidoscope and its infinitive pattern. From this idea we created a sort of hexagonal pattern made of marble. This feeling reminded us a lot of Arabic decoration, especially the Moroccan tiles. The floor has a strong impact when you enter the club, and at the same time when you are on the crowded dance floor it represents a surprise when your eyes go down and find out this decoration.

The defining features of this design:
The materials, definitely all combined together are creating the unique image of the DROP. Traditional material such as wood mixed with stainless shiny surfaces is the feature of the bar: this contamination is the real soul of the club.

How this project reflects its location:
The project tries to tell about the story of the area and the building. This building used to be a storage and headquarters for an English company in the 19th century. That's the reason behind the decoration inside. Before our renovation a Chinese restaurant was located here for several years, and the original decorations were hidden in the false ceiling as the owner was not interested at all in showing those off. So once the false ceilings were removed we decided to preserve these decorations, mainly on the columns and on the beams, and let them become one of the key elements of the design. This area of Shanghai has always been in the past one of the most international, as near the river there were the headquarters of foreigner companies. You can see this also in the architectural style of the building. A mix coming out from all the different countries operates in this area.

How designing a project in Shanghai differs to designing projects in other parts of the world:
Shanghai has a deep tradition, history and heritage. Unfortunately often it is not shown enough or it is totally forgotten. Not all the clients are interested in telling about Shanghai and about its story. We were lucky as our client was interested when we mentioned about preserving the building and the space and showing off its own character.

Generally speaking Shanghai is an international city, and only a few projects are really linked to its tradition. The area of the Bund, where also our project is located, is the only one where we can find good examples of projects that have special relations with the history.

The challenges in designing this project:
The challenges were mainly two: one is about the design, and the other about the construction. About the design, as the client knew clearly what kind of feeling he wanted and we had to make him happy and of course to create something outstanding. In a city as Shanghai the club "market" is quite tough and it was important for the DROP to launch itself also with an outstanding design.

Then designing a club where hundreds of people can be hosted is not an

easy task, because you always have to match the design with the functionality of the space.

Also from constructional point of view it was a big challenge. Sometimes in Asia the quality of construction is not that high. Luckily we worked both with Italian and local contractors, all very experienced and the result was great. Sculpture-like elements such as the bar and the DJ booth, or the decoration of the VIP rooms, to mention some, could result terrible if the contractors were not skilled enough.

How these challenges were overcome: Honestly, with the design part we never had problems, as the client always accepted our ideas. We had a clear mutual understanding since the beginning. From the constructional point of view every time we had some specific requests, both client and contractors supported us, as they perfectly understood the importance of the quality for such an important project.

永恒的空间设计
——欧比可建筑设计

设计师对上海DROP俱乐部的简单介绍：
客户要求我们创建一个现代经典主义的室内设计。一个既要流行又不要太时髦的室内设计。太过时髦的空间很快就会过时，我们的客户想要创建一个可以长期保持其形象的俱乐部。因此，我们从现代经典主义的角度开始设计整个室内。

项目细节描述：
一个"现代经典"设计引导我们对家具陈高的第一选择：俱乐部内所有的座位都使用了切斯特菲尔德沙发。这是一个经典设计，却又带有现代主义鲜亮的色彩，给人一种欧式的感觉。另外一个非常重要的选择是有关墙面的。细木护壁板非常的传统，但又与墙面分开，像是浮动的元素。也使人有欧式的感觉。然后是地板，地板上的装饰也成了俱乐部里其他包装和图案的创意。设计地板的时候，我们受到了万花筒及其千变万化的形式的启发。利用这个创意，我们设计了一种用大理石做成的六边形。这种感觉让我们想起了许多阿拉伯装饰，尤其是摩洛哥瓷砖。当你进入俱乐部的时候，地板会给你很强的震撼，同时当你踏入拥挤的舞池低头看到这个装饰的时候，它会给你以惊喜。

设计的典型特征：
材料，当然所有的元素结合在一起创造出了DROP独特的形象。传统的材料如木材加上酒吧内不锈钢的泛光墙面：这种组合是俱乐部真正的灵魂。

这个项目的设计是怎样体现其地理位置的：
此项目试图向大家讲述其地理位置与建筑本身之间的关系。这座建筑在19世纪的时候曾是一间仓库和一家英国公司的总部。这就是内部装饰产生的原因。在我们翻新之前，这里有一家开了很多年的中餐馆。最初的装饰被掩藏在临时的天花板里，因为主人并不想将这些装饰显露出来。因此，当临时天花板被拆除之后，我们决定保留这些装饰，主要是柱子和横梁上的装饰，并让这些装饰成为设计的关键元素。这个地点一直以来都是上海最国际化的部分，因为与水相邻，许多外企的总部都坐落在这里。通过建筑物的建筑风格，你也可以看出这一点。来自不同国家的建筑风格融合在了一起。

在上海设计一个项目与在其他国家进行设计的不同之处：
上海有深远的传统、历史和文化遗产。可惜的是它经常被掩盖或者完全被遗忘。不是所有的客户都有兴趣讲述上海及上海的故事。我们很幸运，因为当我们提及保留建筑和空间，展示其特征的时候，我们的客户对此很感兴趣。

一般来说，上海是一个国际性城市，且仅有少数项目真正与其传统相关。我们项目所处的Bund区，是能找到与历史有特殊关联的项目的唯一区域。

设计此项目所遇到的挑战：
挑战主要有两个：一个是设计，另外一个是建造。关于设计，因为客户明确知道他想要什么感觉，所以我们一定要让他满意，而且当然要创造出一些杰出的东西。在像上海这样的城市里，俱乐部的"市场"很艰难，所以对于DROP俱乐部来说，杰出的设计是非常重要的。

设计一个可以接待上百人的俱乐部并不容易，因为你不得不将空间的功能性与设计相融合。从建筑的角度来看，这也是一个巨大的挑战。有时候亚洲建筑的质量并不是很高。很幸运的是，我们与意大利及当地的承包商共同合作，他们都经验丰富，因此最终的效果很好。如果承包商技术不高超的话，雕塑式元素，如酒吧和DJ台，或者贵宾室中的装饰会非常糟糕。

困难是怎样被克服的：
老实说，设计方面我们从未有过问题，因为客户总是接受我们的想法。从一开始我们就互相理解。从建筑的角度来说，每当我们有一些特殊要求的时候，客户和承包商都很支持我们，因为他们非常清楚质量对于这样一个重要项目的重要性。

1. Bright view of the bar
2. Red leather sofas increase the dignity of the VIP room
3. Bright bar counter with various bar benches
4. The long leather sofa and simple bar benches provide different seating
5. Light in and above the bar counter shining together
6. Different seating areas create different atmospheres
7. Red sofas display the elegant character of the club

1. 亮丽的吧台设计
2. 红色皮沙发增加了贵宾室的高贵气质
3. 耀眼的吧台及不同形式的吧凳
4. 长长的皮质沙发和简约的吧凳为顾客提供了不同的就座环境
5. 吧台内部及上方的灯光交相辉映
6. 不同的就座区形成了不同的氛围
7. 红色沙发展示了俱乐部高雅的特点

and classic leather sofa, so-called Chesterfield, became the outstanding seating of DROP. That was the tradition, but designers also have to mix it with the modernity.

For the bar and DJ booth, there are two sculptures of stainless steel in champagne finishing, where the three-dimensional decorations create vibrant and warm surfaces that will constantly reflect different tones of light. The space is organised on different levels, with the VIP rooms and the seating areas located on platforms. In this way the space is not flat, and different areas where you can see and you can be seen are created. The whole club is presenting the same floor finishing all around the areas: a geometrical Moroccan tile became the motif of the precious marble floor. This warm decoration becomes another element that shows clearly the crossover of the interior design of DROP: elements of different cultures and times connected together in order to create a unique atmosphere, an atmosphere that has to be sophisticated and easygoing at the same time.

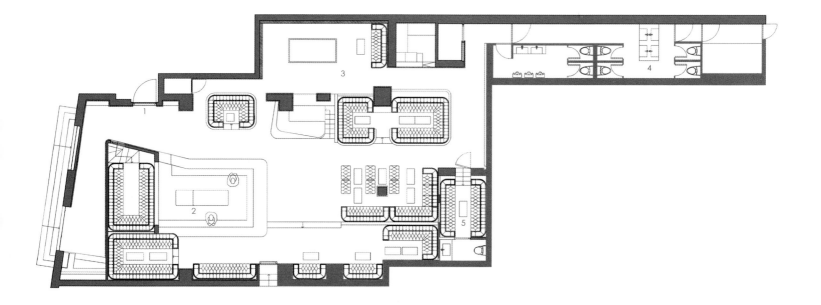

1. Entrance   1. 入口
2. Bar counter   2. 吧台
3. VIP room 1   3. 贵宾室1
4. Restroom   4. 卫生间
5. VIP room 2   5. 贵宾室2

4

5

About the Sunny Blue Sea Club
– By Fong Tsun

1. Brief introduction: The Sunny Blue Sea Club is a clubhouse containing bath, massage, and entertainment, located in Chengdu, China, owned by the Sunny Blue Sea Group. It is a public clubhouse opened to all but most of its space is for members only.

2. Design style: ART DECO style

3. Design concept:

Since this is the second time of decoration, the design focuses on functions combining styles to achieve to be practical and comfortable.

A) Functions feature for the first: Take advantage of the existing structures, minimise the change of the existing building and maximise the visual reconstruction.

B) Use of ART DECO style: Combine various elements, updated traditional elements and spaces to obtain a geometric, totem-exaggerated, bright and colourful but cultural space. From carpets to sofas and decorative pictures, all of these elements are telling their own stories.

C) Based on the practical use of the space, the design combines with environmental protection, health and sustainability to choose every detailed material, so as to maximise the comfort of the space.

From the design of Sunny Blue Sea Club, we could see the states of contemporary clubhouse.

Generally speaking in terms of scope of use, there are two types of club: A

type: community club, which exists in the commercial or residential communities in cities, providing basic services of fitness and pleasure; B type: public clubs, including commercial clubs and holiday clubs, spreading in cities and various resorts, providing high-quality services of fitness and social parties. Such clubs are of successful operating models. Briefly speaking, all the past clubs focus on providing services for fitness and pleasure, including the erstwhile Western clubs and native China clubs. The contemporary clubs are more places for leisure, fitness, especially for parties for friends. If we define home as space of family love, then the club today should be defined as space for friendship.

关于海兰晴天会所
——方峻

1.简介：海兰晴天会所是海兰晴天集团旗下位于中国.成都的一所集洗浴按摩、娱乐于一体的休闲会所。其面向所有人开放，但绝大部分采用会员制的形式存在的公共会所。

2.设计风格：新装饰主义(ART DECO)风格。

3.设计理念：

由于是二次装修，所以在设计时，以功能为主，风格为辅，做到实用与舒适并行。

A.功能为先。根据原有的空间建筑构成特点因势利导，做到最小的原建筑改动，最大化的空间视觉元素重构。

B.新装饰主义风格的运用。把各种时尚元素、更新的传统元素与空间结合在一起，便得到一个几何化、图腾夸大化，有艳丽的色彩，但却还是显得很有文化涵养，从地毯到沙发、饰画都在述说着自己的故事！

C.从实用出发，兼顾环保、健康和可持续。精心挑选每一种材料，使空间的舒适度最大化。

透过海兰晴天会所，看今天的会所业态。

从使用的范围上总述，当今的会所不外乎两类：A类，社区会所，存在于现在很多城市的各大商业非商业居住社区，为社区居民提供基本的健身、休闲娱乐服务；B类，公共会所，包括商业会所和度假会所，广布于城市与各风景区，提供相对高端优质的休闲健身与聚会服务，拥有成功的商业运作模式。简言之，都是以提供身心健康的休闲娱乐服务为主，包括往昔西方的会所或者中国土生土长的会馆。现在的会馆更多的是作为一个休闲健身、娱乐聚会的友情据点形式而存在、发展着。那么，如果把家定义成亲情的据点，则今天的会所便可定义为友情的据点。

Sunny Blue Sea Club

海兰晴天休闲会所

Location:
Chengdu, China

Designer:
Hong Kong Fong Wong Architects & Associates

Photographer:
Wang Jianlin

Area:
3,000 m²

Completion date:
2007

项目地点：
中国，成都

设计师：
香港方黄建筑师事务所

摄影师：
王剑林

面积：
3000平方米

完成时间：
2007

The designer combines neoclassicism with modern design skills. The solemn, sacred and permanent shaping and compositing skill was combined with traditional elements, components and modern installation technology, making classic walk with time. The overall layout is arranged through spatial sequence of symmetry, modest and cumulative progression in classical architectures. The details are not created according to the erstwhile architectural elements. They are created in a typological way, simplifying the classical architectural patterns into triangles, columns, cuboids, blocks and so on, so as to express the abstract classical feelings and the inner link. The interior design focuses on using furniture and artworks to highlight the historical and classical characteristics.

新古典主义设计语言同现代设计手法相结合，将古典中庄严、肃穆、神圣、永久的造型和构图手法，传统要素、元件和现代的安装工艺结合在一起，使古典与时间同行。大型构图上采用古典建筑的对称、稳重、层层递进的空间序列。细部上却不采用往昔的建筑元素，而是以类型学的方式，将古典建筑样式极度简化为三角体、圆柱体、长方体、契块等，从而表现抽象的古典意味及内在的神似。在室内安装时，着重了应用家具和陈列艺术品来增强历史古典特色。

2

Interactive Space
– By Frenjick Quesada

Club Defined
A club in the modern setting is a place where people with a common thread come together to celebrate life. The place and the people reflect each other. Each feeds into the atmosphere and energy of the event. This symbiosis comes alive as the space and the inhabitants synergise.

Design Philosophy
Design HQ believes in listening to the client. While we design predominantly in the modern and eclectic styles, we recognise each client is unique in style preference and functionality requirement. The result of each project is truly a collaboration between Design HQ and the client. We continually strive for a fresh look with an eye for detail reflecting the individuality of the client.

Designing Club Ascend
In designing Club Ascend, our approach was to capture its differentiating essence from the other clubs in the city. With the name Ascend, the design team chose to anchor its design on modern abstract forms accentuating upward and dynamic movement. This was captured in details such as an unfolding origami ceiling, swirling circular banquet seating, jagged intersecting lines and sinuous biomorphic forms. Details in the interior such as ascending small to-large-scale chairs and textures that played well with dramatic lights reinforced the design.

Space allocation was a challenge since we had three areas, the restaurant lounge, the club proper and the VIP section that had to be distinct but had to be able to interact with one another. Locating them strategically adjacent or one on top of the other allowed this. Another challenge was balancing both flexible and built-in modules to attain the target seating capacity. A combination of linear and curved seating forms allowed a good juxtaposition for a range of groups of customers. Modules that could be easily repositioned and moved around allowed this. Lighting and acoustics was key in the design. Being a venue for evening events, the colour palette and material selection had to be chosen in this context. Thus, more pronounced and bold textures in fabric, forms and material were selected since they had to stand out in the context of a darker yet active lighting scheme. Going easy on designs in some areas where we

knew specialty lighting would be used allowed the light effects to be better appreciated. We were fortunate to know upfront the sound system specifications and equipment that would be used. The proper placement and design treatment allowed the design to work well with this important feature of the club.

Designing Club Ascend was an all-around challenge. Balancing function and aesthetic and working closely with client and contractors and suppliers, we were able to collaborate well to capture our joint vision while building on each other's ideas.

互动空间
——弗兰吉克·克萨达

俱乐部的定义
现代背景下的俱乐部指的是享有共同爱好的人们聚在一起庆祝生活的地方。俱乐部与人们互相呼应。每个人都能感受到这里的氛围和能量。当空间与人们互相融入活动的盛况中时，这种互利关系就更加明显。

设计理念
HQ设计事务所相信客户的意见。当我们的设计主要以现代兼收并蓄的形式展开的时候，我们意识到每个客户对风格的偏好和功能的需求都是独一无二的。每个项目都是HQ设计事务所与客户共同合作的结果。我们不断的追求新鲜的设计和细节以反映每个客户的独特之处。

攀升俱乐部的设计过程
在设计攀升俱乐部的过程中，我们的方案是要把握其与这个城市中其他俱乐部的不同之处。因其名为"攀升"俱乐部，设计组选择将其设计成现代抽象的形式以突出向上和流动的动态感觉。这个概念在细节上可以得到体现，例如一个伸展开的折纸形天花板、漩涡式盘旋的宴会座椅、锯齿形相交的线条和弯曲的生物形态的结构。室内的装饰细节，如攀升的由小到大的椅子与动态的灯光相配合，突出了设计的特点。

空间分配对设计来说是一项挑战，因为我们有三个区域要设计：餐厅休闲区、俱乐部和贵宾区，这些区域都要彼此分开但又必须要互相配合。只能采用相邻设置或者上下叠层的形式。另外一项挑战是要平衡具有流动性的座椅且要按比例安置以达到理想的室内容量。直线与曲线排列方式相结合使得一定范围内的顾客可以舒适地就座。这一点是通过可以轻易移动的座椅模式来实现的。照明和声音是设计中的关键部分。作为晚间盛事的举办场所，色调和材料一定要根据环境背景来选择。因此更多的选择了结构、形式和质地上显著大胆的材料，因为这些材料一定要在较暗的背景及灵动的灯光下引人注目。我们知道有些地方需要使用特殊的照明设置，这使得灯光效果更好。幸运的是我们预先就了解了音响系统的规格和所要使用的设备。适当的安置和设计处理使得其与俱乐部的特点配合得天衣无缝。

设计攀升俱乐部是一项巨大的挑战。平衡功能与美学上的关系，与客户、承包商和供应商紧密合作，使得我们可以把我们共同的理念很好的体现在设计上。

Club Ascend

攀升俱乐部

Location:
Metro Manila, The Philippines

Designer:
Collaboration of Design Hirayama +
Quesada and G.R. Flancia & Associates

Photographer:
Erik Liongoren

Area:
1,000 m²

Completion date:
2007

项目地点：
菲律宾，马尼拉

设计师：
Hirayama + Quesada与G.R. Flancia &
Associates联合设计

摄影师：
埃里克·莱昂格仁

面积：
1000平方米

完成时间：
2007

Club Ascend, with over 1,000 square metres of space, is a premier, high-energy dance club, restaurant and cocktail lounge in Metro Manila. One of the unique features of Ascend is that each area has different functions or moods. The challenge was for each area to have its distinct look and yet blend into each other seamlessly. There are three areas that can be enjoyed separately and can be unified to be one big party. The layout was designed with this in mind. The club as the nucleus of activity can be viewed from the overhead VIP balcony. A sliding glass divider allows the restaurant to be part of the club if the event requires it. Functionally, the design had to balance addressing operational needs of maximising seating in defined but interactive restaurant, VIP, and dance club areas.

The design played with dramatic origami shapes, bold active swirls and provocative modern forms. This aims to give a sense of movement, dynamism and layering inspired by the club's name – Ascend.

Key design features would be an abstract origami accent ceiling treatment to counterpoint ascending chairs upon entering the restaurant. Semi-circular swirling seating further emphasises the active design. This is all set in a sophisticated dramatic dark colour palette. Upon entering the club, jagged lines and bold forms accentuate the dance floor and DJ booth

空间面积1000多平方米的攀升俱乐部是一所位于马尼拉的顶级舞蹈俱乐部、餐厅和鸡尾酒休闲吧。

攀升俱乐部的一个独到之处在于每个区域都有不同的功能或氛围。最大的挑战就是要让每个区域都有其独特的外观，且又要彼此天一无缝的融合在一起。共有三个区域可供独立享受且又可以合成一个大的聚会场所。整体布局设计始终贯穿着这个思想。人们可以从上方的贵宾厅看到俱乐部这个所有活动的核心区域。当需要的时候，可滑动的玻璃护栏可以使餐厅成为俱乐部的一部分。从功能上来讲，设计一定要平衡经营对就座率最大化的需求，又要保证餐厅、贵宾厅和舞厅空间的互动。

此设计采用了引人注目的手工折纸的形状、积极大胆的漩涡式和浓烈的现代主义形式，以俱乐部的名字"攀升"为启发，旨在给人以动态、活力和层次感。

最关键的设计特征是在一进入餐厅处采用了抽象的手工折纸式的天花板与攀升的椅子相呼应。半圆形漩涡式座位进一步突出了设计的活跃性。所有这些装饰都是以厚重的引人注目的深色为背景。一进入俱乐部，锯齿状的线条和大胆的风格突出了中心位置的舞池和DJ台。过渡到顶层的专属区所采用的则是丰富而细致的结构来欢迎顾客的光临。对于那些想放松一下、呼吸一下新鲜空气的顾客，可以从贵宾区进入到一个清新整洁，结构自然有序的顶层室外露台。在这个有利的位置上，顾客可以欣赏到马尼拉天空的美丽轮廓。

1. The VIP Lounge mixes biomorphic shapes with origami forms and active lines
2. The sculptural origami ceiling echoes in the lounge
3. Layers of lighting and different levels of seating in the lounge
4. Movable layered tables in the club area can be reconfigured to accommodate different event requirements
5. A sliding door allows the lounge to be separate from the club area but easily accessible
6. Bright lighting with comfortable sofas provide fabulous atmosphere
7. The elegant colour adds dignity to the VIP lounge

1. 贵宾室内有生物形态的设计、手工折纸式的装饰和生动的线条
2. 手工折纸式的天花板笼罩着休闲厅
3. 休闲吧内不同层面的灯光及就座区设计
4. 可移动的桌椅可以改装，为举办不同的活动提供了便利
5. 拉门将休闲室与俱乐部区域分开，人们又可进出自如
6. 明亮的灯光和舒适的沙发为人们提供了极好的氛围
7. 典雅的色调增加了贵宾室的高贵气质

in the centre. Transitioning to the top-floor exclusive area, rich and detailed textures are introduced to welcome the guests. For guests wanting a breath of fresh air, a pristine and immaculate outdoor top-floor balcony with natural and organic textures is accessible from the VIP area. A view of the Manila skyline is enjoyed from this vantage point.

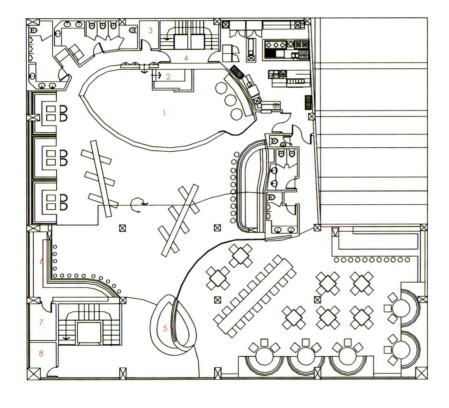

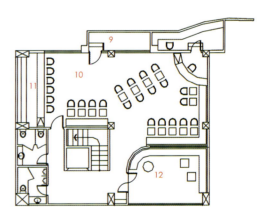

1. Dance floor stage
2. DJ's booth
3. Dressing room
4. First exit
5. Reception
6. Bar
7. Office
8. Storage room
9. Balcony
10. VIP lounge
11. Bar
12. Outdoor lounge/smoking section

1. 舞池
2. DJ厅
3. 化妆间
4. 第一出口
5. 接待处
6. 吧台
7. 办公室
8. 存储室
9. 阳台
10. 贵宾休息室
11. 吧台
12. 室外休息室–吸烟区

4

5

6

7

Exclusive Restroom Design of the Cocoon Club
– By 3deluxe

In design terms the sanitary facilities of the Cocoon Club are reduced to the bare essentials, creating a counterpoint to the opulence of the club and restaurant interiors. The almost archaic sense to the sanitary zone is primarily the result of the anthracite grey colour and the rough texture of the wall tiles, which are top-to-bottom.

White sidelights from wrap-around lighting coves accentuate the natural stone structure of the randomly arranged tiles. This irregular arrangement adds a lively note to the walls, while the vertical orientation of the narrow tiles seems to increase the room's height. A seamless black terrazzo floor covering cast in one piece forms a stark colour contrast to the light grey plastered ceiling.

White ceramic rectangular washbasins are installed in a row on a masonry plinth. Brushed stainless steel soap and towel dispensers set into the wall are located between every two of them. Large mirrors form a continuous wall-to-wall glass surface that is illuminated from above by single-strip light-emitting diffuse light. At every washbasin there is an atmospheric video image projected by the 6" LCD monitors concealed behind two-way mirrors. This technical trick creates a fascinating projection effect as the image appears to be directly superimposed onto the mirror glass.

The pinkish-violet toilet partitions stand out both as regards colour and design. A photoprinted motif conceived by 3deluxe graphics embellishes their thermopal surface. The pattern is composed of pine branches mirrored several times. Depending on how far away the observer stands, the pattern is perceived differently, kindling a variety of associations. Seen from a distance the greenery has the abstract look of a Rorschach inkblot, while from closer up it is easier to spot its affinity to the Asian-style cherry blossom motif featured in the adjoining InBetween Lounge. To ensure the continuity of the pattern, the cubicle doors are fitted with special hinges that cause them to close automatically, preventing any break in the image caused by a door left open.

By contrast, the purely functional fittings – such as the built-in cupboard covered in black linoleum in the men's toilets – take a discreet back seat.

Moreover, every effort was made to incorporate technical equipment as unobtrusively as possible: ventilation is via ceiling slits; and from behind these same slits fluorescent lamps emit golden white indirect light. Eighteen mid- and high-range loudspeakers set into the suspended ceiling together with another eight subwoofers ensure the balanced presence of the music from the Cocoon Club Soundweb.

In the overall architectural design less was also more – an ingenious spatial arrangement made it possible to forgo the use of doors between the corridor area, the lobby and the toilet zone. The resulting uninterrupted space allows for the unimpeded movement of persons even in times of high frequentation.

蚕茧俱乐部卫生间的独特设计
——3deluxe 设计公司

从设计的角度来讲，蚕茧俱乐部里的卫生设施被缩减到只保留重要的部分，这与俱乐部和餐厅室内的豪华形成了鲜明对比。几乎过时的卫生区从上到下铺满了煤灰色的粗糙的墙砖。

环绕式灯槽里安有白色侧灯。突出了随意摆放的瓷砖的自然石结构。这种不规则摆放给墙壁增添了生动的气息，按垂直方向排列的窄砖似乎又增加了室内的高度。地面上铺设了一整块无缝的水磨石地板，与浅灰色的石膏天花板形成了鲜明对比。

白色矩形陶瓷脸盆排成一行安在砖石堆砌成的底座上。每两个脸盆之间的墙上都安有不锈钢的肥皂架和毛巾架。上方灯带发出来的四散光照在墙面的大镜子上，照亮了整个墙面。每个脸盆处都有一个视频图像，这是由隐藏在双向镜面后面的6寸液晶显示器所投射出来的影像。当图像重叠地投射在镜子上的时候，这个技术上的小窍门产生了迷人的投射效果。

略带粉紫色的卫生间隔断无论从颜色上还是设计上都很出色。3deluxe公司设计出的影印图案修饰了隔断的表面。整个图案由反射了许多次的松枝图案组成。根据人们所站位置的远近，人们所感受到的图案也有所不同，激起了人们的无限的联想。从远处看，枝叶图案好像抽象的罗夏墨迹，而从近处看又很容易将其与相邻休闲吧内亚洲风格的樱花图案相联系。为了确保图案的连续性，室内的门使用了特殊的转轴，使门可以自动关闭，这样可以避免产生图案的不完整状态。

相比之下我们可以看到，功能性十足的家具反而变成了不起眼的陪衬——比如男卫生间墙中镶嵌的柜橱，上面覆盖着黑色的油毯。

另外，设计师想尽一切办法使所加入的技术设备不显眼：换气通风是通过天花板的缝隙进行的；透过这些缝隙，荧光灯从后面发出金白色的光。18个中高音扬声器和8个亚低音扬声器镶嵌在悬浮的天花板上，以确保来自蚕茧俱乐部声音系统中的音乐可以均衡的传出。

在整个建筑设计过程中，少即是多：巧妙的空间设计使得走廊、厅堂与卫生间之间不必安装门。所产生的连续空间使人们即使在高度繁忙的时段也可以自由通行、畅通无阻。

Cocoon Club

蚕茧俱乐部

Location:
Frankfurt, Germany

Designer:
3deluxe

Photographer:
Emanuel Raab

Area:
2,700 m²

Completion date:
2004

项目地点：
德国，法兰克福

设计师：
3deluxe

摄影师：
伊曼纽尔·拉伯

面积：
2700平方米

完成时间：
2004

The Cocoon Club was conceived as an avant-garde field of experimentation where space and perception could be transformed. What makes the concept for this forward-looking club and its two restaurants so special is its constantly changing semi-virtual atmosphere. Located at various heights, thirteen capsule-shaped, glazed micro-rooms penetrate the membrane wall, enabling eye contact between the quieter outer area and the buzzing interior. They are modern loges, providing privacy for "cocooning" in the midst of a semi-public environment.

The Cocoon Club is a three-dimensional interface; the entire space is like an instrument, which can be played in acoustic and visual terms by DJ and RoomJockey (VJ).

The perforated walls of the triangular main section are reminiscent of a permeable cell membrane, whose openings permit the flow of guests between the surrounding walk-about area, the "InBetween", and the dance area. Thanks to several layers of white flowstone modules and the resulting deep texture of the outer surface, the so-called "membrane wall" is one of the club's most significant architectural elements. A moving 360° projection, which can be synchronised in time to suit the DJ's set, is projected onto the side facing the dance area. The fact that this digital layer fits exactly over the membrane wall creates an impression

蚕茧俱乐部被视为实验空间与感觉转换的先锋。使这个前卫的俱乐部和其两个餐厅如此特别的原因是其不断改变的半虚拟的氛围。位于不同高度的13个胶囊状、像玻璃一样光滑的微型小屋有着薄膜墙面，使安静的外区与嘈杂的内区有交流的机会。这种现代式包厢，为身处半公共环境中的人们提供了不受干扰的环境。

蚕茧俱乐部是一个立体的界面；整个空间就像一个乐器，可以被DJ或者节目主持人所利用以提供听觉和视觉上的表演。

三角形主区域的带孔墙壁使人们想起可渗透性的细胞膜，打开的时候，可以接纳在四周闲逛的以及舞池里的客人。因为有了若干个白色浮石模块层和由此而产生的外部空间表面的深色材质，所谓的"薄膜墙"成为了整个俱乐部中最重要的建筑元素。一个可以360度移动的投影仪投射到舞池对面，可以同一时间配合DJ的背景。事实上，这个电子层与薄膜墙之间相互配合，产生了一种生动的印象——它的材质似乎都被溶解了。

这个570平方米的舞池被深色木制平台分隔开。弯曲平台的亮边在立体的地板上形成了装饰图案。平台共分五个层次，包括一个台阶、2个为舞蹈表演者准备的圆柱、一个贵宾休闲区和3个酒吧。

蚕茧俱乐部里最突出的元素无疑是这庄严的白色DJ操作台，其有机形态就像与薄膜墙壁一同生长出来的一样。DJ和主持人在这个凸出来的位置上控制多媒体装置，因而控制舞池中每个瞬间的气氛。使用专门定制的软件来操作多媒体技术，为控制俱乐部的氛围提供了无限的机会。

1. Three bar areas are located on platforms on the dance floor
2. The uninterrupted space allows for the unimpeded movement of people even in times of high frequentation
3. The organic shape of the DJ pulpit seems to have grown together with the membrane wall
4. The membrane wall is the most significant architectonic element of the club
5. The Cocoons provide privacy in the immediate vicinity of the dance floor
6. "InBetween" is the atmospheric mood zone

1.三个酒吧区位于舞池的平台上
2.连续的空间使人们即使在繁忙时段也可以畅通无阻
3.器官形状的DJ台就像与薄膜墙同时生长出来的一样
4.薄膜墙是俱乐部内最重要的建筑元素
5.蚕茧结构营造了邻近舞池的私密空间
6."间隙"空间氛围迥然

of liveliness – its material nature appears to be dissolving.

The 570-square-metre dance floor is divided up by a landscape of dark wood platforms. The illuminated edges of the curved platforms form an ornamental pattern on the three-dimensional floor. The platforms are at five different levels and include a stage, two columns for dance performances, a VIP Lounge and three bars.

The most striking element in the Cocoon Club is undoubtedly the imposing white DJ pulpit, whose organic form seems to have grown together with the membrane wall. From their raised position, DJ and RoomJockey conduct the multi-media staging, thus controlling the momentary mood on the dance floor. The media technology, operated using specially customised software, provides limitless opportunities for manipulating the atmosphere in the club.

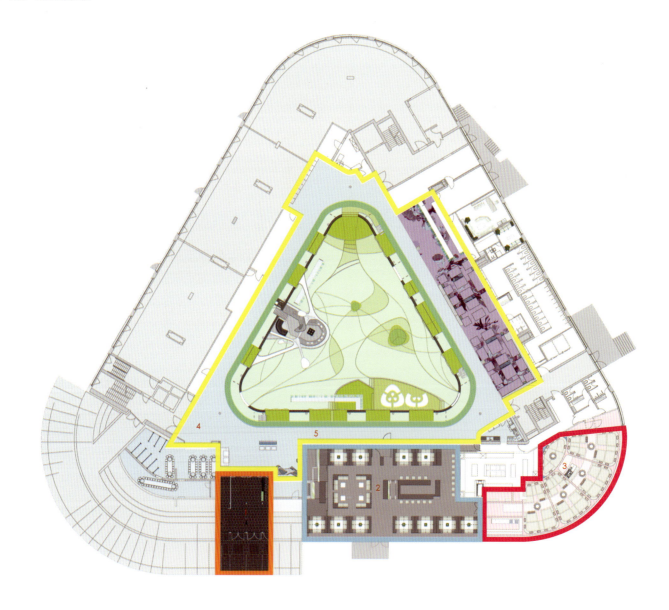

1. Entrance
2. Micro club restaurant
3. Silk bed restaurant
4. InBetween with lounge
5. Membrane wall

1.入口
2.微型俱乐部餐厅
3.丝绸餐厅
4.间隙空间&休闲区
5.薄膜墙

4

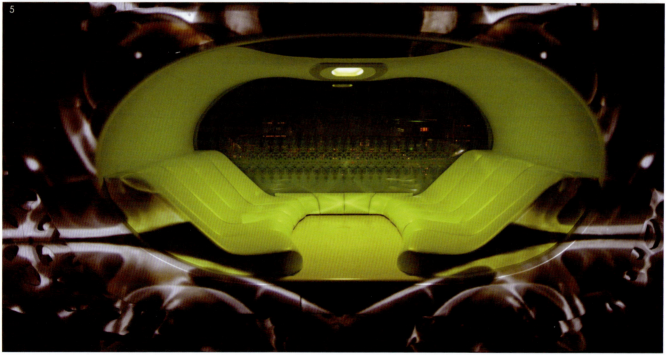

5

The Design Concept of the Champagne Lounge
– By Marshall Kusinski Design Consultants

The concept for the Must Champagne Bar and Lounge was to imagine and create a space which would incite a sense of luxury, decadence, mystery and whimsy. A space very different from in atmosphere to MUST, yet still connected as a secretive, mysterious VIP location – to be discovered or to be inviting to... The signature or hero of this space is the product – champagne, and of course therefore the space captures some of the allures and delights of champagne itself.

In subtle and dramatic ways champagne has been translated into the feel and look of the space... through materials that glisten, shimmer and bubble to lighting that highlights and softens. The colour-and-material palette has been selected to be coherent and glossy with reflection and shine of importance. This, together with features in textured and heavily patterned and detailed fabrics contrasts to enhance the space and create this dramatic effect of "golden" "glowing" and stands out within the deep glossy black of the walls, ceilings and floors around you. The main colours will therefore be black in a myriad of finishes, together with opulent features in gold and yellow to reinforce the idea of champagne...

Firstly the only marker from downstairs where there is another space to discover is the door; this will be a glossy black door with a hint to what's

beyond... Once invited to join the champagne experience, clients will enter via a new staircase to the upstairs bar. This staircase will be fairly minimal so as not to "give away" the surprise above, yet a striking glowing wall constructed from "caviar glass" in a champagne colour will capture people's excitement and thirst as they walk up the stairs viewing only glimpses of the dark space above.

The main bar and lounge area is one open room, like a parlour, boudoir or salon, with a variety and choice of seating options clustered together to form three intimate separate groups or one space where people can choose their favourite spot to lounge, sit, drink, sample and chat in a space which is exciting and comfortable.

There is an element of whimsy too, with one wall being dedicated to items or a collection of mirrors, pictures, found and vintage objects or pieces of whimsy such as a gold elk head.

Plush velvety drapes adjacent to black glass add depth, drama and reflection on one side of the space reflecting the golden glow of the long champagne display. This linear display is designed as a showcase which leads your eye to the main bar at the end of the space. This bar, highlighting champagne as the "hero" of the space is the highest element in the space playing and drawing your eyes up into the height

of the loft ceiling above and making a feature of the delicate yet dramatic feature pendants.

Three vintage, lampshade-inspired tassel pendants from Belgium feature in the lounge and are designed to subtly distinguish the three seating clusters, whilst adding drama and intrigue. The bar is designed to be open, so clientele can interact with the whole experience – the creation of a sumptuous cocktail or the art and care taken with a special champagne...

An exciting and unusual element to this bar is a radical champagne display located in the centre of the bar. Your bottle of champagne will sit open and on display with ice building up around it, keeping it chilled whilst drawing attention, with the ice creating a misty effect. The theatre of serving champagne is highlighted through this important element.

The seating is a collection of chaise style long lounges, where people can relax, sit close, and stretch-out. Along with Louis, French vintage inspired single armchairs and parlour chairs. These will all be in a selection of decadent and detailed fabrics to add variety and richness to the space. Clusters of lounges and chairs are teamed with glossy reflective crystal-like cubes used to rest a drink on and individual table lamps to add golden yellow light and warmth. The space

can be transformed into a private dining space for 14, in a formal long table setting with crystal-like-Victorian inspired chairs to add interest and keep within the delicate reflectivity of the material palette. Furniture and extra storage are practically located behind a concealed door for staff to use with ease – whilst retaining the polished and reflective wall panelling without compromise.

Ambient lighting, combined with reflective, dark, dramatic surroundings works as a backdrop for the gold and detailed furniture and whimsical features whilst allowing the product – champagne – and other top-shelf drinks to remain the feature and retain maximum impact in the lounge where wonderful champagne is the reason you are there experiencing the space and the excitement it inspires.

香槟休闲吧设计理念
——Marshall Kusinski设计顾问公司

玛斯特休闲吧的设计理念是要想象并创造一个空间，可以激起人们奢华、神秘而奇妙的感觉。一个与玛斯特酒吧气氛完全不同，且作为有创造性的神秘贵宾场所与酒吧有某种联系的一个空间。这个空间的特征就是其产品——香槟酒，因此这个空间捕捉到了香槟酒本身的一些魅力及其所带来的愉悦。

香槟酒以精致美妙的方式通过闪光的材料和突出而柔和的照明设计融入了空间之中。精选的色调和材料相一致并闪耀着华美的光泽，与厚重质感的图案和细致的纹理相对比，加强了空间的"金灿灿""光闪闪"的效果，且在周围的黑色泛光的墙面、天花板和地板的衬托下显得很突出。无数装饰的主色调都是黑色，其他奢华装饰为金色和黄色——更加强化了香槟休闲吧的设计理念……

楼下唯一可以探索的标志就是门；将设计一扇黑色泛光的门以暗示楼上更深层的神秘……一旦前来体验香槟酒的乐趣，客人们便会经过崭新的楼梯来到楼上的酒吧。楼梯非常小，这样不会提前泄露楼上的惊喜。当客人们走上楼梯，瞥见楼上昏暗的空间里，由"鱼卵式玻璃"筑成的香槟酒颜色的墙将会激起人们的兴奋和渴望。

休闲吧主区域是一间开放的房间，就像一间客厅、闺房或者沙龙，有很多座位可以选择，这些座位分成了3组，人们可以选择最喜爱的地方来休息、就座、喝酒、聊天，感受兴奋和舒适。

这里也有怪异的元素：由镜子、画及其他奇异的物品如金麋鹿头装饰的墙面。

奢华的天鹅绒幕帘与相邻的黑色玻璃增加了空间的深度及强烈的效果——展现长长的香槟酒陈列台所反射出的金色光芒。流线型的陈列台被设计成了玻璃陈列柜的样式，将您的视线引导到空间尽头处的酒吧区。酒吧以香槟酒为其"英雄"，是这个空间内的最高元素，将您的视线指引到上面的天花板，突出了其精美而引人注目的垂饰。

3个优美的灯罩配上比利时风格的流苏垂饰巧妙地将休闲吧内的三个就座区域区分开来，且又增加了精致复杂的效果。酒吧设计要让顾客与整个环境产生互动——举办带有特殊香槟酒的奢华鸡尾酒会或者艺术节等。

酒吧内令人兴奋、非同寻常的元素是位于酒吧中央的香槟酒展示柜。打开的香槟酒放在里面，周围铺满冰块，既引人注目又使酒保持冰凉的状态，冰块能营造朦胧的意境。香槟服务台在这个重要元素的衬托下显得更加耀目。

室内的座位都是长长的躺椅，在这里客人们可以放松、靠近坐在一起、或者伸展全身。还有法国路易风格的扶手椅和会客椅。所有这些椅子上都配有精心挑选的精致织物，增加了空间的多样性和丰富性。长椅、短椅与用来放饮品的水晶式立方台相结合，桌灯散发出金黄色的光芒，营造了温馨的氛围。这个空间可以转化成可容纳14个人的私人餐厅，正式的长桌，似水晶的维多利亚式椅子，在精致高雅中不失趣味性。器具和另外的存储间都安排在一个隐蔽的门后面，工作人员可以安心使用——同时又有使泛光的墙面保持光泽。

周围的灯光与反光的暗色调环境一起为金色家具、古怪的装饰以及香槟酒和其他饮品充当了背景。极妙的香槟酒便是你来到这里体验空间及兴奋感的真正原因。

1. Long lounges and single armchairs are teamed together
2. The crystal-like chairs and formal dining table match perfectly together
3. The champagne display shelf shining with the crystal-like chairs
4. Table lamps give off warm light
5. The long and soft chaise style benches provide comfortable experience
6. The champagne is the "hero" of the lounge

1. 长形休闲椅与单个的扶手椅相互呼应
2. 水晶般的椅子与餐桌完美结合
3. 香槟酒柜与水晶般的椅子交相辉映
4. 台灯散发出温和的光
5. 长而柔软的贵妃椅为顾客提供了舒适的享受
6. 香槟才是俱乐部真正的"英雄"

"It is time to provide a serious Champagne experience in a lush and opulent space. You'll feel spoilt, privileged and excited just to be there. Luxury, privacy and comfort are what the lounge is all about," says Russell.

In subtle and dramatic ways Champagne has been translated into the feel and look of the space from the materials that glisten, shimmer and bubble to lighting that highlights and softens. Tracey Bear, Interior Designer from Marshall Kusinski Design Consultants has risen to the challenge of designing a space that can also be transformed into a private dining room for 14. The lounges disappear behind cleverly concealed storage and a formal long table setting with crystal-like Victorian-inspired chairs (in keeping with the delicate reflectivity of the material palette) creates a beautiful setting.

The Champagne Lounge is open every evening – seven days – and is available for private lunches, and may be booked for private evening events as well. MKDC won a Design Institute of Australia (DIA) Award of Merit in 2008 in the division "Interior Spaces – Professional", in recognition of outstanding design qualities achieved.

香槟休闲酒吧一周7天每天晚上都开放。客人可以来这里享用私人午餐，也可以提前预定夜晚的私人专场。MKDC获得了2008年澳大利亚设计协会"专业组室内空间"优秀奖，以表彰其设计质量上的突出成就。

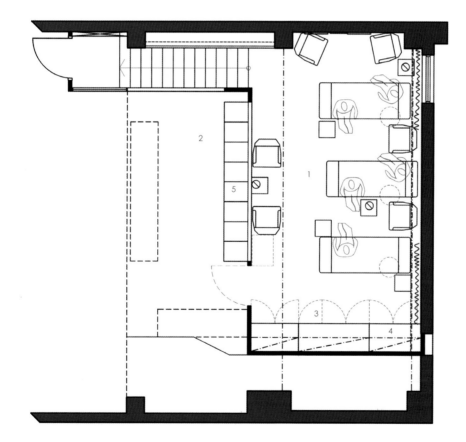

1. Lounge/Bar
2. Store area
3. Bar
4. Service area
5. Champagne display fridges

1. 休闲吧
2. 存储区
3. 酒吧
4. 服务区
5. 香槟展示柜

The Pacific Club
太平洋俱乐部

Location:
Newport Beach, CA, USA

Designer:
MVE & Partners

Photographer:
Eric Figge

Area:
2,787 m²

Completion date:
2006

项目地点：
美国，加利福尼亚州，新港滩

设计师：
MVE及其合伙人公司

摄影师：
埃里克·菲格

面积：
2787平方米

完成时间：
2006

An exclusive private membership club, The Pacific Club's original structure was completely demolished and rebuilt in 11 months on this prestigious site that includes a peaceful lake in the heart of Newport Beach, CA. The 2,787-square-metre project was three years in the planning stages. Mediterranean-inspired design influences the building exterior, interior, entry tower and porte cochere of this world-class fine dining, urban rendezvous.

Bringing quality and detail together, Robert Puleo, Director of Interiors and Associate Partner for the architecture firm MVE & Partners, combined traditional design motifs with a strong axial plan. Showcasing impressive art treasures, the interior architecture frames the gallery hall, library, meeting rooms, ballroom, dining rooms, grill and wine cellar. Nautical watercolours, California plein air paintings, modern works of art, and oils from the Joan Irvine Collection enhance every setting.

A timeless, comfortable environment blends formality with the Newport Beach lifestyle. Elegant granite flooring in the lobby leads to a library warmed by a Puleo-designed fireplace and hand-milled American walnut panelling that is used throughout the club. On axis with the library is the living room, designed as a comfortable gathering space enhanced with a limestone fireplace and walnut flooring.

作为一间奢华的私人会员制俱乐部，在11个月内，太平洋俱乐部的原始建筑被彻底拆除并重建于这个享有胜名的地点。这里有位于美国新港滩中心的一个宁静的湖泊。这个2787平方米的项目设计历时3年。地中海风格影响了这座世界顶级的就餐聚会建筑的外部设计、室内设计、入口处塔的设计和门廊设计。

室内设计主管及MVE及其合伙人公司的中级合伙人罗伯特·溥乐将质量和细节紧密联系在一起，融合了传统的设计图案和强劲的对称设计。构建画廊、图书馆、会客厅、舞厅、餐厅、烧烤厅和酒窖的室内结构充分展示了引人注目的艺术财富。航海水彩画、加州外光画、现代艺术作品及琼·欧文的油画收藏品都各个设计背景增添了韵味。

永恒而舒适的环境将新港滩的生活方式融入到传统礼节中。前厅高雅的花岗岩地板通向图书馆，里面有溥乐设计的壁炉和手工制作的美国胡桃木镶板，这种镶板贯穿了俱乐部的整个设计。客厅与图书馆处在同一轴线上，作为舒适的聚会场所，里面有石灰岩壁炉和胡桃木地板，加强了舒适的效果。

在整个俱乐部中，大小不同的私人餐厅为就餐与会议提供了不同的场所。主舞厅内配备了精致的手工吹制的穆拉诺玻璃吊灯。为了提供更加随意的环境，还专门设计了塔尔博特烧烤厅和酒吧，里面的铜墙纸和压花皮面板为室内增添了生趣。在烧烤厅里，有艺术感的开放式厨房强化了人们的美食体验，通过一排法式门可以到达就餐露台，在这里可以俯瞰科尔湖。

酒窖餐厅重塑了旧世界的氛围，砖地板、手工涂制的灰泥墙和一个穹窿形拱顶支撑着传统熟铁锻造的吊灯。透过玻璃墙，客人们可以直接看到受气候控制的酒窖。

1. The fireplace of the library could be seen from the lobby
2. The floor, wall and planter match with each other
3. The fireplace in the library creates warm experience
4. The sofas in the living room create a comfortable atmosphere
5. Exquisite ceiling and neat design create fabulous dining experience
6. Different sofas in the lounge bar provide different atmospheres
7. Brick flooring creates a traditional feeling

1.从前厅可以看到图书室的壁炉
2.地板、墙面和花盆和谐统一
3.图书室的壁炉为顾客提供了温馨的体验
4.起居室的沙发营造了舒适的氛围
5.精美的天花板和整洁的设计营造了美妙的就餐环境
6.休闲吧内不同的沙发营造了不同的氛围
7.砖地板营造了传统的感觉

Throughout the club, private dining rooms of varying sizes offer opportunities for dining and meeting in different settings. The main ballroom features an exquisite hand-blown Murano glass chandelier. For more casual ambiance, there is the Talbot Grill and adjacent bar, where copper wallcovering and embossed leather panels add interest. In the Grill, a state-of-the-art display kitchen heightens the culinary experience, while sets of French doors open up to a dining terrace overlooking Lake Koll.

The wine cellar dining room recreates an Old World ambiance with brick flooring, hand-troweled plaster walls and a groin vault supporting a custom wrought-iron chandelier. A glass wall affords diners a direct view into the climate-controlled wine cellar.

1. Main entrance	1.主入口
2. Vestibule	2.门廊
3. Lobby	3.前厅
4. Library	4.图书室
5. The living room	5.客厅
6. Main dining room	6.主餐厅
7. Grill dining room	7.烧烤餐厅
8. Dining terrace	8.就餐露台
9. Main kitchen	9.主厨房
10. Members' bar	10.会员吧
11. Wine cellar	11.酒窖
12. Private dining	12.餐厅包房
13. Existing spa&fitness centre	13.原有水疗&健身中心
14. Existing bistro&terrace	14.原有酒吧&露台
15. Renovated men's locker room	15.翻新男士衣帽间

4

5

Polo Club
马球俱乐部

Location:
Berkshire, UK

Designer:
Casa Forma

Photographer:
Barry Murphy

Area:
372 m²

Completion date:
2008

项目地点：
英国，伯克郡

设计师：
Casa Forma设计事务所

摄影师：
巴利·梅菲

面积：
372平方米

完成时间：
2008

The brief consisted of developing an extension to the historical building to create a clubhouse to include a bar area, a dining room, a cinema room, a conservatory, a large kitchen, a guest accommodation suite and an adjacent wing to house the offices and meeting rooms for the management of the club. From the very beginning the horses were also a prime consideration as the existing stables weren't at the standards to accommodate such valuable thoroughbred horses. The landscape would also need to be manicured and redesigned but always with the intention of keeping natural and harmonic with the strong character of the English country life.

It would need to be smart and reflect the lifestyle of an authentic English heritage. Naturally, horses and polo activities represent masculinity and power. The designer took these concepts on board to select natural stones, reclaimed red bricks, slate tiles for flooring and cladding surfaces. The new building was partially clad with bespoke oak panelling to warm the rooms and add interest to the walls creating a sophisticated backdrop for the selection of Ralph Lauren furniture.

The layout was designed carefully to make the seating areas integrated to the conservatory and bar area. The main focal point was the central freestanding fire place clad with solid oak mantelpiece and the chimney was clad

简要来说，这个项目是要扩建这座历史建筑，创建一个内含酒吧、餐厅、放映厅、暖房、大型厨房及住宿套间的俱乐部建筑，且要在其相邻位置安置办公室和会议室，以便管理俱乐部之用。从最开始的时候，马也是这个项目首要考虑的对象，因为现存的马房无法达到安置如此珍贵的纯种马的标准。这里的景观也需要修整和重新设计，但要始终遵循一个原则就是要保持其自然性，使之与浓烈的英国乡村生活特征和谐一致。

设计要巧妙地反映出纯正的英国传统生活方式。自然地，马和马球运动象征着刚毅和力量。设计师根据这些理念选择天然石材、再生红砖、石板岩来铺设地面。新建筑有些部分使用了特制的橡木镶板，既保温又给墙面增加了生趣，为所使用的拉尔夫·劳伦家具创造了很高雅的背景。

布局设计很仔细，使就座区与暖房和酒吧区相融合。主要的焦点是中心独立式的壁炉，镶嵌在实心的橡木壁炉架里，烟囱上镶有粗糙的铜板，产生了石板墙的效果。酒吧内特制的家具和墙板都衬有巧克力棕色皮革，上面印有鳄鱼标记。采用灯心绒、仿古皮革、山羊绒和格伦花格呢羊毛等材料来复现英国乡村风格。新建筑与旧建筑等高，但所有的新山墙都显露在外以保持天花板的高度。山墙全部采用的是涂蜡的橡木材料，全部手工磨制，以配合设计理念。

为了扩大俱乐部内的社交区域，景观的很大一部分被再生石板所覆盖以创建一个露台，上面摆设了大桌子，为室外活动提供场所。在露台的位置可以俯瞰马房和现存的钟塔。由于重新设计的马房屋顶比原来的屋顶高了3米，为了使钟塔的高度与之保持相应的比例，设计师们使用再生红砖将钟塔加高了。重新设计马房是为了提供更好的马室、更多的空气流通和更好的照明，以改善在比赛期间所伺养的50匹马的生活质量。

1. Guests could enjoy the warm sunshine in the conservatory
2. The candleholder on the table is atmospheric
3. Spacious stable
4. The lounge design is very elegant
5. The design of guest suite is of a rustic atmosphere
6. Comfortable seating and exclusive decorations fulfill the space

1. 客人们在暖房里可以享受温暖的阳光
2. 桌上摆放的烛台别具韵味
3. 宽敞的马房
4. 休闲室设计非常典雅高贵
5. 客房设计带有乡村田园气息
6. 舒适的座椅与独特的装饰点缀了空间

with a rustic copper slate creating an effect of stack stone wall. The bespoke bar furniture and wall panels were lined with chocolate brown leather with alligator print. Textures such as corduroy, distressed leather, cashmere and a selection of glen plaid wools were used to bring the English country style to live. The new section of the building followed the height of the existing building, while internally all the new gables were kept exposed to provide ceiling height. The gables were all made of solid waxed finish oak, all hand distressed to add character to the design concept.

In order to expand the internal social area of the clubhouse, a large section of the landscape was clad with reclaimed flagstones to create a patio with large tables for outside events. The patio was located overlooking the stables and the existing clock tower. The clock

tower was raised using reclaimed red brick in order to be proportional to the new heights of the stables which had new roofs completely designed and raised by three metres higher than the original ones. The stables were redesigned to provide better bays and more ventilation and better light to improve the quality of life for the 50 horses accommodated during the polo season.

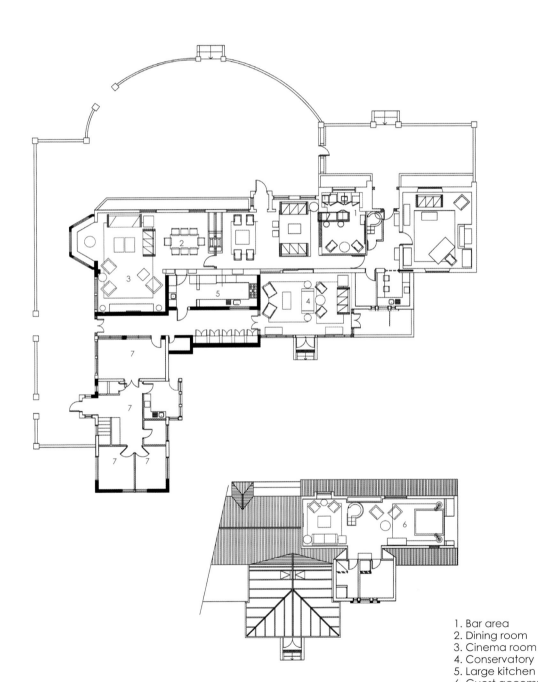

1. Bar area
2. Dining room
3. Cinema room
4. Conservatory
5. Large kitchen
6. Guest accommodation suite
7. Offices and meeting rooms

1. 酒吧区
2. 餐厅
3. 放映室
4. 暖房
5. 大型厨房
6. 客房
7. 办公室和会议室

Golf Club House
高尔夫俱乐部

Location:
Domat / Ems, Switzerland

Designer:
Karsten Schmidt - IDA 14

Photographer:
Andrea Flak

Area:
840 m²

Completion date:
2002

项目地点：
瑞士，多玛特/埃姆斯

设计师：
卡斯滕·施密特—IDA14 设计公司

摄影师：
安德莉娅·弗莱克

面积：
840平方米

完成时间：
2002

The modern glass architecture of Andrea Gubbini of the golf club building Domat / Ems was with an interior design concept that provides the high design standards of the building needs and a hospitable atmosphere.

In the whole restaurant area, the designers have used stonewall coverings seperated by wooden semi-transparent seperations which were designed to both seperate different types of seating units and get a wide point of view with the graphical holes in them. These decorative holes allowed the desingers to use some of the wooden elements as shelves. Floors and 13-metre-long bar were covered with natural stone. This bar held many functions inside such as storage of cutlery, plates and bottled drinks, music system and cashiers desk, not to forget small patisserie products offered to customers to taste. The lay-out plan was formed by using free tables along entrance axis, a long social table, small tables of different sizes and shapes which could be combined and seperated according to the number of customer groups. Also, some compartments like private seating areas were formed between wooden seperations.

Different usage of the tables was one of the most important considerations for the owners, so the designers have placed most of the small tables in front of the long banquette seatings. Ceilings around the area were lowered on

Andrea Gubbini的多玛特埃姆斯高尔夫俱乐部建筑是一座现代的玻璃建筑，其室内设计理念满足了其高标准的建筑要求，为其营造了热情好客的氛围。

在整个餐厅区，设计师都采用了石墙涂层，由木制的半透明的隔板隔开，这样既将不同形式的座位分开，又能让顾客通过隔板的图案孔看到外面的广阔空间。因为有了这些装饰性的图案孔，使得设计师可以采用木制元素做成架子。地板和13米长的吧台都铺设了天然石材。吧台区有很多用途，可以存放刀具碗碟、瓶装饮品和音响系统；充当收银台；还可以提供法式蛋糕等供客人品尝。室内布局为：沿着入口处摆放了一些免费桌椅，一张长的社交用桌，还有一些不同大小和形状的桌子，可以根据客人的多少进行组合或者分开摆放。利用木制的隔断，还形成了一些像隔间一样的私人就座区。

桌子的不同用途对于所有者来说是最重要的。因此，设计师将大多数小桌子摆放在了长条软座位的前面。座位上方的天花板高度被调低，营造一种舒适感。天花板上面覆盖着木板。还选择了一些吊灯装饰，为夜晚就餐的人们提供淡淡的灯光。天花板的其他部分则相反，吊得很高，颜色很浅，使空间显得很宽敞。木制隔断和墙面上都挂有蔬果素描，为整个空间增添了一丝幽默感。

厨房区的封闭部分被建筑的中层一分为二，形成一个员工衣帽间和一个冷藏间。还可以在这个封闭区洗涤碗碟，存放烹饪用具。

1. Dining tables combine perfectly with the ceiling
2. Interior in the light of "blue aura"
3. The red wall is a great background for "still life"
4. Toilet design shows simplicity
5. Different seating areas provide different experience
6. The light and bar area shining together
7. The furniture in the Member Lounge is simple but comfortable

1.餐桌与天花板完美结合
2."蓝光"下的室内
3.红色的墙面很好的衬托了这"静静的生命"
4.简约的卫生间设计
5.不同的座位为会员提供了不同的体验
6.灯光与酒吧区交相辉映
7.会员休闲室内陈设简约却舒适

top of the built-in seating to create a sense of cosiness and they were covered by wooden panels, and some suspended lighting fixtures were chosen to have a dim light for late night dinners. The other parts of the ceilings, on the contrary, were kept high and light-coloured to help keep the spacious feeling. Wooden seperations and wall coverings were supported by sketch images of vegetables to add a little sense of humour to the whole place.

The closed part of the kitchen area was multiplied by two by building a mezzanine which made it possible for the designers to create locker rooms for the crowded staff and cold rooms as well. In the closed part of the kitchen, scullery and cooking devices were placed.

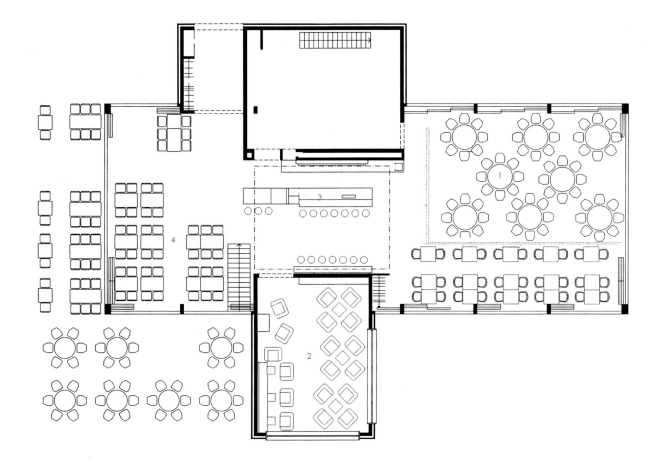

1. Restaurant   1.餐厅
2. Member Lounge   2.会员休闲吧
3. Bar   3.酒吧
4. Open-space dining   4.公共就餐区

Foxtail Supperclub
福克斯泰尔俱乐部

Location:
West Hollywood, CA, USA

Designer:
Franklin Studios, Inc.

Photographer:
Grey Crawford, Witt Preston

Area:
365 m²

Completion date:
2008

项目地点：
美国，加利福尼亚州，西好莱坞

设计师：
富兰克林设计有限公司

摄影师：
格雷·克劳福德，威特·普雷斯顿

面积：
365平方米

完成时间：
2008

Foxtail was conceived of as a place where its larger-than-life Hollywood investors could have some drinks, do some work, and bring their celebrity friends. The idea was that the creators of today's Hollywood blockbuster were lacking something – they had the money and the fame, but what they needed to complete the picture was a hangout. Harking back to a more formal time, when stars strove to be glamourous and American prosperity was expressed in its lavish architecture, Art Deco and Art Nouveau styling were used to create an intimate, rich and often surprising atmosphere. Like the movies that might be made over its tables, Foxtail's rich environment has the effect of transporting its guests away from their every-day lives.

The predominant patterns and details have Art Deco and Art Nouveau origins, but have been modified and blended together with an eye on the seventies disco scene. The classic, white-plaster fluting of Art Deco was re-imagined as a glossy black skin on the walls, accented by copper stripes. The classic dome supported by columns is another Art Deco move that was tweaked by making the dome segmented, imprinting it with a Bibaesque relief pattern and illuminating it with discoesque LED lights and runners. The back bar has the ordered, controlled patterning of Art Deco, but is made of frosted glass and is lit from behind. The bathrooms for both sexes are saturated in

福克斯泰尔俱乐部被认为是其具有传奇色彩的好莱坞投资者们饮酒、工作和招待朋友的地方。这个项目的主要思想是：创造现代好莱坞盛事的创造者们缺少某些东西——他们有钱有名，但他们需要放松娱乐来完成他们的人生图画。这可以追溯到从前那些时光：当明星们努力奋斗想要绽放光芒的时候，美国的繁荣通过其奢华的建筑来表现的时候，艺术装饰和新艺术主义风格被用来营造亲密、丰富而惊喜的氛围的时候。就像电影源于生活却高于生活一样，福克斯泰尔俱乐部丰富的环境也可以将客人们带离日常生活的烦杂。

设计的主要形式和细节都有着艺术装饰和新艺术主义的源头，现已经过修改并融入了20世纪70年代迪斯科的韵味。艺术装饰中的经典白石膏凹槽装饰被修改，变成了光滑的黑色墙面，并镶有铜条装饰，更加突出了其特点。艺术装饰中另外一个元素——由圆柱支撑的经典圆顶——也被改动，分成了几个部分，上面印有比芭风格的图案，还有迪斯科式LED灯，灯光闪烁。后面的酒吧配有艺术装饰中有序的图案，但是是由磨砂玻璃制成的，而且光是从背后射出的。男女卫生间都装饰了粉色墙纸，上面印有精致的花纹，圆润的边角，鱼头形水龙头，这些都展示了艺术装饰和新艺术主义中所表现的大胆而柔性的气质。

吸烟休息室内也采用了当年经典的绿色，但是却是在丰富而近乎工业化的铜色墙纸衬托下带有的铜绿色。这种绿色给空间提供了很好的背景，20世纪20年代的枝形吊灯装饰也同样展现在这个空间里。设计师选择了新艺术主义风格，是因为它具有异域风情又有些颓废派的韵味，不拘谨。选择艺术装饰风格是因为其将俱乐部与洛杉矶城市联系在一起，让人们记起这座城市被誉为高雅之城的那个时代。

1. The light in the dining room creates a romantic atmosphere
2. The exclusive design of the ceiling above the dance floor
3. The chandelier is shining like stars
4. Elegant patterns in the space
5. Classic green colour in the smoking lounge
6. Exquisite design of the bar counter

1.餐厅的灯光营造了浪漫的气氛
2.舞池上方的天花板设计独特
3.枝形吊灯如繁星般闪耀
4.空间内的图案设计典雅
5.吸烟休息室内的经典绿色
6.吧台设计精美

pink wall covering with a subtle floral pattern, rounded corners and fish-head faucets, giving a nod to the boldl feminine qualities expressed in Art Deco and Art Nouveau.

The smoking lounge has also used the classic green of the time, but finds the colour in a rich, almost industrial copper wallpaper with a strong patina. This green provides a nice backdrop for the vintage, the 1920's deco chandeliers hung in the same room. The desingers chose Art Nouveau because it is exotic and decadent without being stuffy. They chose Art Deco because it connects the venue to the City of Los Angeles, and reminds people of a time when this city was still considered elegant.

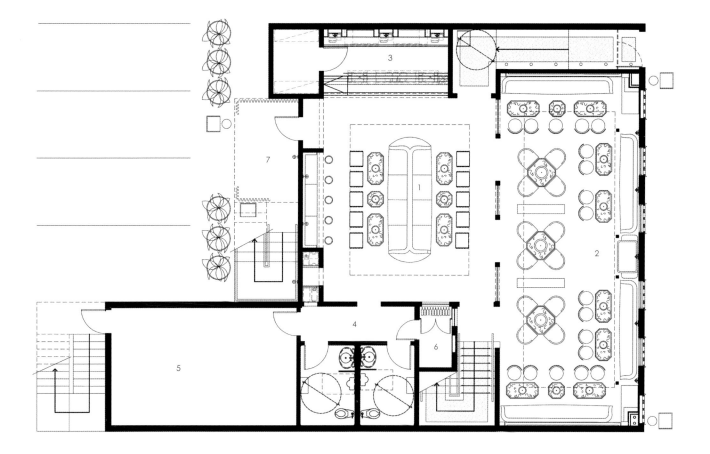

1. Lounge   1.休闲吧
2. Dining room   2.餐厅
3. Bar   3.酒吧
4. Bathrooms   4.浴室
5. Kitchen   5.厨房
6. DJ booth   6.DJ厅
7. Entrance   7.入口

1

Beijing Whampoa Club
北京黄埔俱乐部

Location:
Beijing, China

Designer:
Lyndon Neri and Rossana Hu

Photographer:
Derryck Menere

Area:
2,080 m²

Completion date:
2007

项目地点：
中国，北京

设计师：
郭锡恩，胡如珊——如恩设计研究室

摄影师：
Derryck Menere

面积：
2080平方米

完成时间：
2007

ENTRANCE/CORRIDORS: Upon entering the restaurant and travelling through the corridors, one is surrounded by a completely white space. Solemn and serene, the purity of the space draws attention to the Chinese construction details rather than obscuring it with colourful imagery, as in history. The white corridors provide rest for the eyes before their transition to the various decadent destinations.

BAR: In contrast with the white corridors, the Bar is all black. Here, the traditional Chinese screen has been replaced and re-interpreted with a custom-made pattern – rings of water overlapping one another, drawing in the water from the courtyard.

PRIVATE DINING ROOMS: The Private Dining Rooms are each a different colour, framed by lacquered custom screens. The decadence of Chinese culture is evident with the northernmost room – historically the most auspicious – the most elaborate.

SUNKEN COURTYARD: The project site is actually a courtyard house re-built where one used to stand, with a parking garage trenched below. For this reason the structures are considered "semi-historical". As commentary on this recent demolition trend and disregard for historic preservation, the first courtyard was hollowed out and opened to the level below, resulting in a physical lowering. Glass was then installed

入口/走廊：一进入餐厅，走在走廊之中，人们会被一片全白的空间所包围，庄严而静谧。运用纯洁的空间将人们的视线引到中国式的结构和装饰上，而并非采用五颜六色的装饰让人眼花缭乱。白色的走廊让人们在进入到其他地方之前，视觉上得到片刻的休息。

酒吧：酒吧是全黑色调的，与白色的走廊形成鲜明的对比。在这里，传统的中国屏风被替换成了特别定制的样式——一个个水环彼此相叠，将庭院里的水引入室内。

餐厅包房：餐厅的每个包房都有不同的色调，由喷漆的特制屏风隔开。中国文化的悠久性在最北面的包房里体现得最明显——这是有史以来最宽敞的、最精心布置的包房。

水中庭院：这个项目的地点实际上是一个重建的庭院旧址，在房子下面曾经有一个地下停车场。由于这个原因，这些建筑结构被视为是"半历史性"建筑。由于当下的拆毁旧建筑的趋势及对历史保护的忽视，第一个庭院被挖空，使平面降低了。并安装上了玻璃，里面填满了水，形成了反光水池，这宁静的背景让人们回想起已经消失了的庭院房屋。

气派的正面大楼梯：气派的楼梯陈列橱展示了备餐的过程。沿着楼梯走，人们可以一阶一阶的感受中式菜肴烹饪的漫长过程。当到达下面的餐厅就座之后，人们可以感受食物本身带来的享受。

公共餐厅：灯光通过上面"支离破碎"的庭院洒进来，又经过成千上万个不锈钢片的反射，照亮了公共餐厅空间。

1. Glass bridge connecting indoor & outdoor spaces
2. Red private dining room
3. Private booth
4. View of public dining room
5. Green private dining room
6. The detailed design

1.玻璃桥连接室内与室外空间
2.餐厅包间红厅
3.私人包房
4.公共就餐区
5.餐厅包间绿厅
6.细节设计

to create a reflection pool filled with water, a peaceful setting in memory of courtyard houses long gone.

GRAND STAIRCASE: The Grand Staircase showcases images glorifying the preparation of food. Moving along the staircase, one can experience, stage by stage, the lengthy preparation process involved in Chinese cuisine. Upon arrival below, one is seated to then experience the food itself.

PUBLIC DINING ROOM: Light streams through the "broken" courtyard above and reverberates across thousands of stainless steel discs in the Public Dining Room.

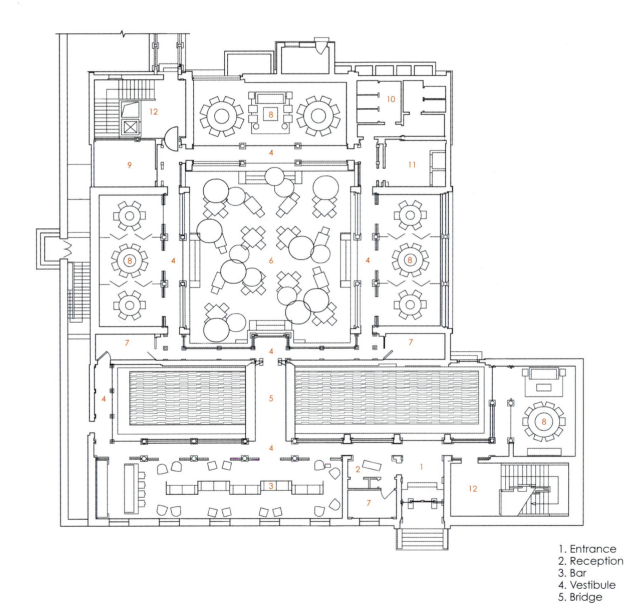

1. Entrance — 1.入口
2. Reception — 2.接待处
3. Bar — 3.酒吧
4. Vestibule — 4.门廊
5. Bridge — 5.桥
6. Courtyard — 6.庭院
7. Storage — 7.存储间
8. Private dining room — 8.餐厅包间
9. Mechanical room — 9.机械室
10. Restroom — 10.卫生间
11. Kitchen area — 11.厨房
12. Grand staircase — 12.大楼梯

4

5

Muse Club

缪斯俱乐部

Location:
Shanghai, China

Designer:
Lyndon Neri and Rossana Hu

Photographer:
Derryck Menere

Area:
1,400 m²

Completion date:
2006

项目地点：
中国，上海

设计师：
郭锡恩，胡如珊——如恩设计研究室

摄影师：
Derryck Menere

面积：
1400平方米

完成时间：
2006

Muse is a club that occupies three different spaces overlooking an atrium in the newly redeveloped Tong Le Fang factories in Shanghai. In a neighbourhood where new nightclubs abound, Muse must differentiate itself by asserting an architectural identity through inventing a spatial experience unlike all the others. The insertion of a cocoon, mimicking the symbol of that womb which transforms the ugly worm into the brilliant butterfly, takes centre stage both physically and metaphysically, serving an array of functions in terms of spatial organisation, yet always reminding people of the transformation that every club-goer secretly desires. Moving bodies and pulsating adrenaline squirm within the ultra-modern/ high-gloss cocoon, where the subtly checkered patterns of white and cream lacquer panels resonate with the lustful eyes that never rest on one single object.

This is where Muse transforms itself into more than a club, but an experience. This element intersects all three functional areas – the dance club, restaurant, and private club – providing a cohesive, unifying object to tie the otherwise dispersed spaces together. Under the hovering cocoon, part of a 20-metre-long bar cast in concrete with bronze edging and clad in green horse hair connects both the restaurant and the club area horizontally. Above the bar, there is a glass structure broken into random segments of vertical glass panels and metal connections

缪斯俱乐部是一家占据三个不同空间的俱乐部，可以俯瞰上海重建的新同乐坊的中庭。在周围都是新夜总会的情况下，缪斯俱乐部必须要通过提供完全不同的空间体验来维持其独特性，这样才能脱颖而出。这里穿插了蚕茧状装饰结构，代表蚕从丑陋的蚕蛹蜕变成美丽的蝴蝶这个过程的发源地，无论从形态上还是感觉上都紧紧与中央舞台联系在一起，既通过宽敞的布局使空间实现了多用性，又总会让人们留意到这种转变，这是每个俱乐部经营者内心都渴望达到的效果。人们在这超现代的蚕茧结构中移动，在这里，白色和奶油色漆板上精致多变的图案随着人们充满渴望的眼神的移动而变化，客人们的眼睛从来都没有一直盯在某一个物体上。

在这里，缪斯俱乐部转变成的不仅仅是一间俱乐部而是一种体验。这种元素贯穿全部的3个功能区——舞厅、餐厅和私人俱乐部——为将其他分散的空间联系在一起提供了统一的目标。在悬浮着的蚕茧结构下面，一个20米长的吧台由混凝土材料构成，上面镶有铜边，吧台台体上覆盖着绿颜色的材料，吧台的一部分水平地将餐厅和俱乐部区域连接在一起。吧台上面有玻璃结构，随意的分成了许多块垂直的玻璃板和金属连接物，将贵宾区和下面的俱乐部垂直地连接在一起。

蚕茧结构后面涂成橘黄色和深棕色的墙上配有抽象的花卉图案，加深了空间的深度、颜色和层次。蚕茧外部的间隙空间形成了很多有趣的背景，这些区域可以作为私人包间、DJ室、就餐室和休息室。

三楼的贵宾区是一个气氛稍缓和的空间，里面配有黑色的仿古家具，同时还配有现代装饰和陈设。三面墙壁上镶有铜、油黑檀木和绿绒装饰，使空间显得朴素低调。

1. The green bar shining with the light
2. The candlelight combines with the light on the ceiling
3. The special design of the ceiling above the bar
4. The neat dining tables and stools
5. Part of the bar connects the restaurant and the club area horizontally
6. Various chairs show the exclusive design idea
7. The glass structure ties the VIP area and the club below vertically

1.绿色的吧台与灯光交相辉映
2.烛光与天花板上的灯光结合在一起
3.吧台上方天花板的独特设计
4.整洁的餐桌与餐凳
5.吧台的一部分将餐厅与俱乐部区域水平地连接 在一起
6.各种各样的椅子显示了独特的设计思想
7.玻璃结构垂直地将贵宾室与下面的俱乐部联系 在一起

that shoot up vertically to tie together both the VIP area and the club down below.

The abstracted floral pattern along the walls painted in orange and dark brown behind the cocoon adds depth, colour, and another layer to the space. Interstitial spaces outside of the cocoon create pockets of interesting settings which serve as private booths, DJ stations, dining banquettes, and restrooms.

The VIP area on the second floor is a more subdued space filled with period furniture painted black and juxtaposed with modern upholstering as well as a collection of modern furnishing. Three walls clad in copper, Macassar Ebony, and green horse hair provide the backdrop for an understated articulation and restrained decadence.

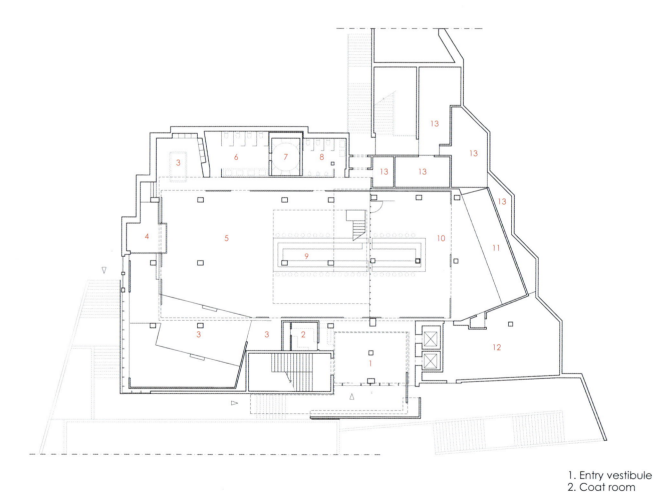

1. Entry vestibule   1.入口门廊
2. Coat room   2.衣帽间
3. Booth   3.大厅
4. DJ booth   4.DJ室
5. Club area   5.俱乐部
6. Women's washroom   6.女洗手间
7. Women's lounge   7.女士休闲吧
8. Men's washroom   8.男洗手间
9. Bar   9.酒吧
10. Dining area   10.就餐区
11. Lounge dining   11.休闲餐厅
12. Kitchen   12.厨房
13. Service area   13.服务区

Neo-Beijing Lounge

新北京画廊

Location:
Beijing, China

Designer:
Kenny Chen

Photographer:
Yang Mingtao

Area:
1,000 m²

Completion date:
2009

项目地点：
中国，北京

设计师：
陈立坚

摄影师：
杨明涛

面积：
1000平方米

完成时间：
2009

The Neo-Beijing Lounge is located in No.1-3 of Nan Xin Granary at No. 22, Eastern Fourty Street of Dongcheng District, Beijing, whose predecessor is an official granary in Ming and Qing dynasty and now is a protected historic building. As the club lounge of CapitaLand auction, it could be "shining" again. Due to its historical values, the designers have to respect and protect its historic trace during the designing process to fully take advantage of the original materials, so as to show its heritage of hundreds of years.

According to the above considerations, the designers used two simple modern materials: steel and glass. They incorporated the two materials together, with proper light textures. The finished space is one that could keep people's steps to appreciate the antiques and to enjoy the elegant atmosphere for reading.

During the process of design and construction, the designers were not allowed to change any original structure of this building, and also not allowed to drill or sprig nails on it. This was a great challenge for both the designers and the builders. While at the same time, this challenge brought the designers much convenience.

新北京画廊现位于北京市东城区东四十条22号南新仓1–3号仓，它的前身是明清两朝的官家粮仓，属于受保护的古建筑群落。现将其作为嘉德拍卖行的会所式酒廊而重新绽放光芒。因其历史价值和时间的沉淀，在设计过程中须尊重与保护其历史痕迹，充分利用原有的一砖一瓦、一木一柱，使其数百年的文化底蕴全方位地再现出来。

基于以上考虑，设计师运用两种简单的现代材料：钢和玻璃。将这两种材料进行碰撞、融合，最后适当地打上灯光。完成后的作品是一处让人愿意停留下来久久回味、细细地赏玩古董和读书的雅境。

在设计和施工过程中，要求设计师对其中的一砖一瓦、一木一柱都不能改变其原有结构，也不能在上面钻孔打钉。由于设计师不能对结构做什么手脚，这无论是对设计还是对施工来说，都是一大难题。但也正是因为这些又给设计师们带来很多便利。

因为这个建筑本身就是一个"美人胚"，不需要怎么花工夫，只需要把它原有的本色和样貌呈现出来就可以了，而呈现一个建筑的原貌，最好而便利的工具莫过于玻璃和灯光，于是这两者便成了这个作品的主角。使这个披着朦胧面纱的美人展现在人们眼前。

1. The rich colours of the bar
2. The parlour is decorated with elegance and luxury
3. The lighting creates a luxurious atmosphere
4. The gorgeous curtains add magnificent quality to the space
5. Classic furniture and patterns create an exclusive space

1.酒吧丰富的色调
2.会客室装饰典雅奢华
3.灯光设计营造了豪华的氛围
4.华美的帷幕为空间增添了高贵的品质
5.经典的陈设和装饰打造了独一无二的空间

Since this building was a "beauty" itself, it need not be decorated so much. It would be great to just show its local features and looks, so as to display the original building. There was no better and more convenient element than glass and light, so the two elements played leading roles in this project. Thus, the veiled beauty showed its face to the public.

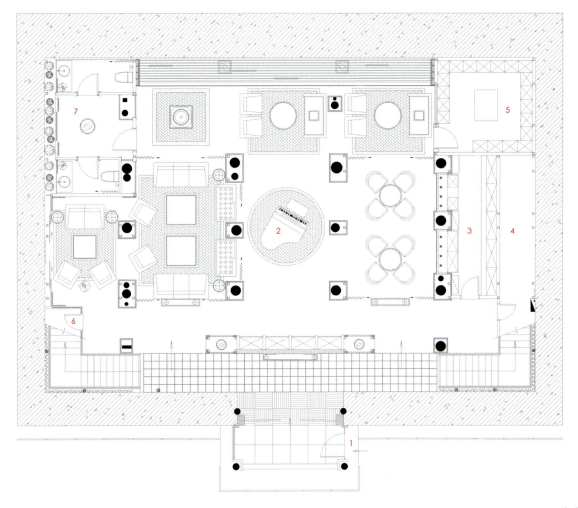

1. Entrance   1.入口
2. Piano   2.钢琴
3. Water bar   3.水吧
4. Work station   4.工作间
5. Wine cellar   5.酒窖
6. DJ booth   6.DJ房
7. Public restroom   7.公共卫生间

The Deuce Lounge

迪尤斯休闲吧

Location:
Las Vegas, USA

Designer:
Franklin Studios, Inc.

Photographer:
Franklin Studios, Inc.

Area:
990 m²

Completion date:
2010

项目地点：
美国，拉斯维加斯

设计师：
富兰克林设计有限公司

摄影师：
富兰克林设计有限公司

面积：
990平方米

完成时间：
2010

MGM Mirage's CityCentre was conceived of as a city within a city. It was to be a place of urban sophistication – a tasteful evolution from the themed casinos that have come to dominate the Las Vegas strip. MGM Mirage sought to replace the theme with high-design in their ten-billion-dollar development, and to this end, hired a collection architects and designers whose work spoke through abstraction rather than direct reference.

Franklin Studios was asked to design a new type of venue – a bar/lounge which incorporates gaming. Guests can simultaneously enjoy the service and atmosphere of an intimate club while enjoying high-stakes gaming. The design was inspired by the surrounding desert with its sweeping geometries and rich textures. For instance, the main walls of the lounge are formed as a lattice of wood, modelled after the cactus skeletons that lie on the desert floor. The function of these open wood walls is to create a sense of enclosure while allowing the energy of the casino and the lounge to pass through and create a visceral connection.

The ceiling is also a light, skeletal form defined by a loose grid of wood members which define a sprawling topography akin to the bare mountains surrounding Las Vegas. The bar front and custom carpet pattern reflect the scalloped forms of wind-blown sand. All of these organic references are formalised and

米高梅公司的CityCentre被认为是城中之城。它是一个充满城市奢华的地方——由主导拉斯维加斯旅游业的主题赌场演变而来的高品位场所。米高梅公司意在使用顶级的设计来塑造这个100亿美元的项目以取代当时的设计主题。为此聘请了一批建筑师和设计大师，这些大师的作品都是抽象的、非直观的。

富兰克林工作室被要求设计一个新型的聚会场所——可以兼有赌博活动的酒吧/休闲吧。客人们可以在享受高风险游戏乐趣的同时享受私人俱乐部式的服务和氛围。此设计受到周围沙漠中包罗万象的几何图形和丰富的纹理的启发。例如，休闲吧的主墙上设计成了木制窗框的造型，模仿的是沙漠上躺着的仙人掌骨架。目的是让这些打开的木墙为客人营造一种被围绕的感觉，而赌场和休闲吧内的热量又可以通过这里进行流通，形成一种内在的联系。

天花板由松散的木制网格构成，轻巧精致，轮廓清晰，就像拉斯维加斯周围光秃秃的山一样。酒吧正面和特制地毯的图案反映的是被风吹过的沙子呈现出的锯齿形。所有这些有机体都被形式化，并包含在现代主义的框架内。休闲吧的整体形式是一个分段的盒子。白天的时候，角落处的大窗户滑开，与赌场的地板直接连接起来，晚上的时候门关上，休闲吧就又成了一间高级夜总会。

1. The wine shelf is shining in the light of chandeliers
2. The wall is modelled after the cactus skeletons
3. The view of the wine shelf from the entrance
4. The ceiling is light and skeletal
5. The gaming tables provide guests with great pleasure
6. Even the restroom contains elements from the desert

1.在灯光的照射下，酒架闪耀着光芒
2.墙面模仿的是仙人掌的骨架的造型
3.从入口处看到的酒架
4.天花板轻巧精致，轮廓清晰
5.游戏桌为客人们提供了极大的乐趣
6.卫生间里的沙漠元素

contained within a framework modernism. The overall form of the lounge is a segmented box. The corners of the box are bent forward and reach out into the surrounding casino floor. The large windows in these corners slide back during the day to create a direct connection with the casino floor, while at night the doors are closed and the lounge becomes an exclusive nightclub.

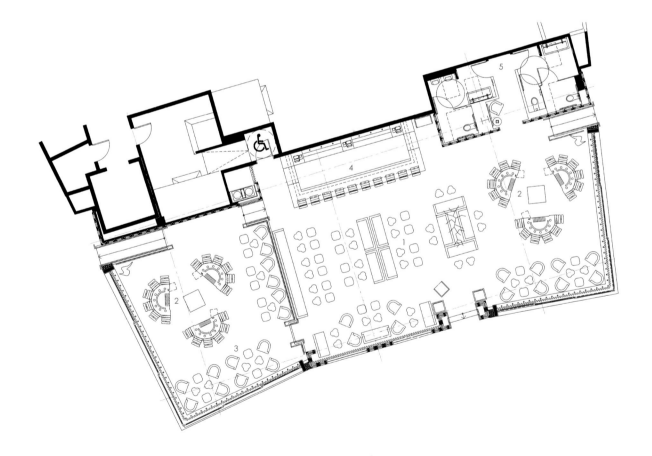

1. Main lounge	1.主休闲吧
2. Gaming tables	2.游戏室
3. Private lounge	3.私人休闲吧
4. Bar	4.酒吧
5. Bathrooms	5.浴室

3

4

5

6

Libre Cigar Lounge Shanghai HILTON HOTEL

上海希尔顿酒店Libre雪茄休闲吧

Location:
Shanghai, China

Designer:
Takeshi Sano / SSDesign Inc.

Photographer:
Nacasa & Partners Inc.

Area:
122 m²

Completion date:
2009

项目地点：
中国，上海

设计师：
武石山诺／SSDesign有限公司

摄影师：
Nacasa及合伙人有限公司

面积：
122平方米

完成时间：
2009

The Libre Cigar Lounge is the social space where people can enjoy Cuban cigar and Cuban spirit. The interior theme is Gentleman's club and Old Shanghai. The key color of this lounge are Gold, Black and Khaki. They make this space not only gentle but also gorgeous. The whole design gives people a sense of luxury.

When guests sit down on the sofa that is decorated by buttons and used to be loved by Count Chesterfield who is the symbol of English gentleman, they can see the light on the ceiling designed based on moon. They can feel at ease on such sofa under the moon-like light and enjoy a time of leisure and relaxation.

Inside wall is surrounded by bricks taking the guests to nostalgic memories like the 1920S Shanghai and unordinary life. Various art works on the walls remind people of the glorious past. Dotted lights and pendant lamps on the ceiling are well-organised. Modern design spice is mixed by R-shaped Humidor, bookshelf and bar area designed with SUS and Glass materials. Shelves around the lounge are convenient for the guests to buy the top-level cigars and things related to cigars.

The new spot was opened in Hilton Shanghai where people can spend great time with high-quality cigar smoke and aroma.

Libre雪茄吧是人们可以享受古巴香烟与烈酒的社交场所。室内设计的主题是：绅士俱乐部与旧上海。休闲吧的主色调是金色、黑色和黄褐色。这些主色调使空间显得温和且华美。整个设计给人一种奢华的感觉。

当客人们在沙发上就座的时候，他们可以看到天花板上根据月亮设计而成的灯饰。沙发上装饰有纽扣，这曾经是英国典型绅士切斯特菲尔德康特的最爱。客人们坐在这样的沙发上享受着月光般灯光的照耀会感觉很舒适。他们可以尽情享受这段轻松的时光。

内墙由砖砌成，让人们回忆起20世纪20年代的旧上海及那段不寻常的生活。墙上各种各样的艺术作品让人们想起辉煌的过去。天花板上星罗棋布的灯排列有序。R形烟草贮藏室、货架和吧台设计中都融合了现代的设计情趣，使用了硅和玻璃材料。休闲吧四周的货架为顾客们购买顶级的雪茄及相关用品提供了便利条件。

新休闲吧就开在上海希尔顿酒店里面，在这里，人们可以享受到高品质雪茄的芳香。

1. Comfortable sofas reflect the light of the space
2. The light on the ceiling is designed based on the moon
3. There are many art works on the brick wall
4. The sofas are decorated with buttons
5. The interior theme is Gentleman's club and Old Shanghai
6. The humidor

1.舒适的沙发反射着灯光
2.天花板上的灯是根据月亮设计的
3.砖墙上有许多艺术作品
4.沙发上有纽扣装饰
5.室内设计的主题是绅士俱乐部与旧上海
6.雪茄贮藏室

1. VIP room   1.贵宾室
2. Cigar locker   2.香烟储存柜
3. Bar counter   3.吧台
4. Lounge   4.休闲吧
5. Item shop&books 5.商店&书屋

Bella Collina

贝拉·科里纳高尔夫俱乐部

Location:
Montverde, Florida, USA

Designer:
Marsh & Associates, Inc.

Photographer:
C.J. Walker

Area:
6,040 m²

Completion date:
2008

项目地点：
美国，佛罗里达州，蒙特沃德

设计师：
马什联合设计有限公司

摄影师：
C.J.沃克

面积：
6040平方米

完成时间：
2008

The clubhouse at Bella Collina provides a spectacular blend of old-world style and abundant services. Its plan offers a variety of spaces and experiences: social and secluded, lively and reserved, opulent and casually elegant, and delivers them in delightful indoor and outdoor locations that rejoice in the surrounding golf course, lakes, vineyards, olive trees and orange groves. Comprised of multiple connected structures set atop rolling hills, its exterior massing suggests a quaint and thriving village in Tuscany.

The entry court serves as village gate, where flower petals float in a sunlit fountain among classical stone columns. Ahead a wide stairway leads to the main village court; a tree-lined reflecting pool highlighting its symmetry, and borrowing from the tranquility of the lake beyond.

To each side of the entry court lie doorways that lead to disparate but equally elegant functions. The full-service spa is to the left, with subtle contemporary styling set against the old-world backdrop. The main level includes lavish salon, hydrotherapy, and steam room facilities around a central relaxation area that is enclosed for solitude but open to the sky above. A chandelier-lit grand staircase leads to massage treatment rooms on the upper level.

贝拉·科里纳高尔夫俱乐部为人们提供了引人入胜的旧世界风情和丰富的服务。其方案为客人提供了各种空间和体验：热闹的或僻静的、热烈的或恬淡的、豪华的或自然高雅的。将客人带入令人愉快的室内和室外场所，让其尽情享受周围高尔夫场地、湖泊、葡萄园、橄榄树和橘树林的优美环境。其室外很多设施都建在连绵的山峰顶上，展现了托斯卡纳区奇特繁荣的村庄形象。

入口庭院相当于村庄的大门，花瓣飘浮在阳光普照的喷泉里，喷泉被经典的石柱包围。前面有一条宽阔的楼梯通向村庄的主庭院；绿树成荫阳光掩映的池塘，有着远处湖泊般的宁静，突显了庭园的对称性。

入口庭院的各个方向都有通道，可以通向位置不同但同样优雅的各个功能区。提供全面服务的SPA位于左边，精致现代风格的装饰与旧世界的背景相呼应。主层主要包括奢华沙龙、水疗和蒸汽房，中间是休息区，室内宁静又可以看到头顶的天空。由枝形吊灯照亮的大楼梯通向上层的按摩室。

SPA对面是餐厅入口，这里是俱乐部社交活动的中心。贝拉·科里纳俱乐部为顾客提供了多种多样的就餐体验，从随意的适合家庭的用餐氛围到正式的私人的用餐环境，应有尽有。所有的就餐区域都与室外露台相连，还有一个巨大的草坪，用来举办招待会和音乐会。回到室内，俱乐部的会员们可以享受他们最喜欢的葡萄美酒，这些美酒都储存在大而深的由砖砌成的拱形酒窖里，锁在他们自己的木制储物柜里。他们还可以进入到室外带遮帘的露台上进行私人活动。

1. Soft leather seating complements the hardwood floors and stone walls in the Member Lounge
2. The dining entrance features a stone tile mosaic floor and a brick groin vaulted ceiling
3. A wispy chandelier hangs from the timber ceiling in the Spa stairway
4. Elegance and warmth greet women golfers in the ladies' locker room
5. An onyx-surfaced exhibition counter allows family members to watch a chef create their meal before their eyes
6. Male golf members can finish their round with a drink in the sumptuous Men's Grill
7. The Family Dining Room provides a casual experience

1.会员吧内柔软的皮沙发与硬木地板和石墙相辅相成
2.餐厅入口有石砖马赛克地板和砖制穹窿拱顶
3.束状吊灯悬挂在通向SPA室的楼梯上方的木制天花板上
4.女高尔夫会员衣帽间气氛温暖典雅
5.家人们可以观看厨师在缟玛瑙表面的展台为他们准备餐点
6.男高尔夫会员可以在奢华的男士餐厅享用饮品
7.家庭餐厅气氛随意

Opposite the spa is the dining entrance, the hub of social activity at the clubhouse. Bella Collina offers multiple dining experiences, from casual and family-friendly to formal and private. Dining areas all connect to outdoor terraces, and there is a grand event lawn for large receptions and concerts. Back inside, club members can retrieve their favourite vintages from personalised wooden lockers in the cavernous brick-arched wine cellar, and retire to a veiled outdoor patio for small private events.

To further convey this historic feel, the clubhouse employs traditional building materials, including stone and brick cladding, weathered stucco, wrought iron and steel, clay tile roofing, hand-painted frescoes, and copper and cast stone trim. The result is an experience of comfort and opulence that appeals to the most discerning guest.

Golf Inc. magazine called the Bella Collina facility "magnificent" in naming it their "Private Clubhouse of the Year" for 2008.

为了更进一步传递俱乐部建筑的历史感，设计中还使用了传统的建筑材料，包括石头和砖、风化灰泥、精炼的钢铁、粘土瓦屋顶、手绘壁画及铜和人造石装饰。这一切都为眼光独到的客人们提供了舒适而丰富的体验。

高尔夫俱乐部有限公司杂志称赞贝拉·科里纳高尔夫俱乐部"华丽"的设施，并给予其"2008年度私人会所"称号。

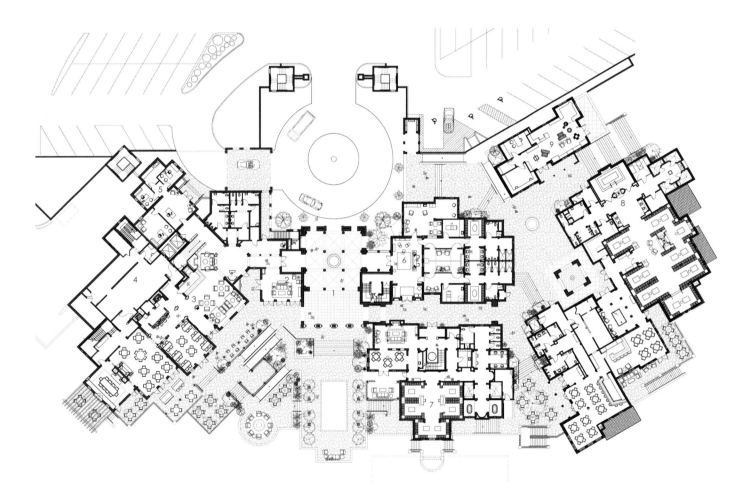

1. Entry court　　　　1.入口中庭
2. Living room　　　　2.起居室
3. Dining　　　　　　3.餐厅
4. Kitchen　　　　　　4.厨房
5. Administration　　　5.行政办公室
6. Spa　　　　　　　6.水疗室
7. Women's lockers　　7.女士衣帽间
8. Men's lockers　　　8.男士衣帽间
9. Golf shop　　　　　9.高尔夫用品商店

1. Sunlight streaming through the glass dome

2. The main Entry Hall resumes the formal symmetry of the exterior, with stone columns, cut limestone floors, and fanciful steel trusses supporting the glass roof

3. Elevated clerestory windows bring daylight to the Men's Locker Room

4. The bar occupies a space with its own pyramid-shaped skylight, dark hardwood floors, and a globe-shaped metal chandelier

5. The Main Dining Room is bright and airy, with floral-patterned carpet, and white painted coffered wood ceilings and wainscot

6. The Men's Lounge evokes an explorer's club

1.阳光从玻璃屋顶照进室内

2.主入口大厅延续着室外的几何造型，有石柱和石灰岩地板，新颖的钢铁架支撑着天花板

3.高高的天窗给男士衣帽间带来了阳光

4.酒吧有其独特的金字塔形的天窗、深色硬木地板和地球仪形状的金属吊灯

5.主餐厅明亮且通风良好，有花形图案的地毯和漆成白色的木制格子天花板及护墙板

6.男士休闲吧犹如探索者俱乐部

the sense of symmetry, formality, and elegance. A succession of gently-pitched stone staircases lead down to the pool, gardens, and greens, providing a graceful transition between building and landscape. The interior palette reflects a British island design, with bright colours, sunlit spaces, and a combination of stone columns, wood wainscots, slowly turning ceiling fans, and furniture and fabrics that suggest a relaxed life of privilege in the Spice Islands. Even the artwork evokes scenes of the British explorer, with shadow boxes containing exotic flowers and insects, and grand globes for recounting tales of adventure and conquest in distant lands.

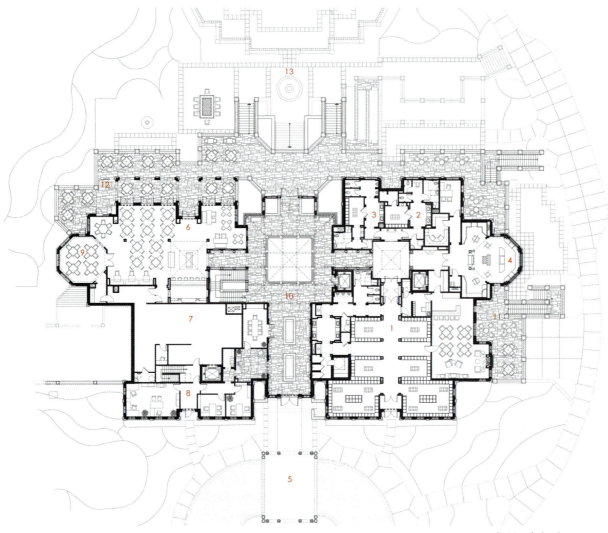

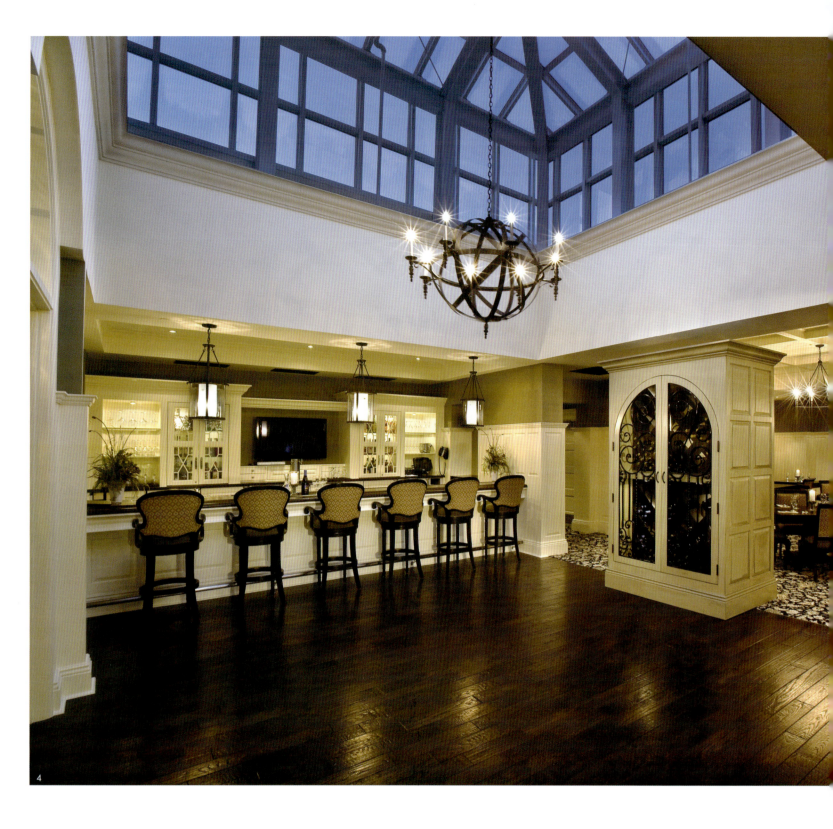

4

Cornerstone Golf Club
奠基石高尔夫俱乐部

Location:
Montrose, Colorado, USA

Designer:
Marsh & Associates, Inc.

Photographer:
Marsh & Associates, Inc.

Area:
600 m²

Completion date:
2008

项目地点：
美国，科罗拉多州，蒙特罗斯市

设计师：
马什联合设计有限公司

摄影师：
马什联合设计有限公司

面积：
600平方米

完成时间：
2008

Thirty-two kilometres southwest of Montrose, Colorado, this 24-square-kilometre parcel of land brimming with mountain wildflowers and old-growth aspen trees lies in view of three ranges of the Rocky Mountains. Here the smell of wild sage fills the crisp mountain air. The clubhouse for the Cornerstone Golf Club conveys the feel of a Mountain Cabin befitting the rugged beauty of this site. This modest structure at the base of a hill was to serve as a temporary facility, but its rustic charm continues to impress visitors more than two years after completion. The building is clad with log siding, and capped with a rusted corrugated steel roof terminating in worn barnwood gables and timber frames. Pilasters made from local Telluride sandstone provide a sturdy base, and foreshadow the central fireplace whose immense stone chimney projects from the building's ridgeline. Inside, pale hand-scraped hardwood floors offset the dark-stained casework and chiseled timber beams overhead. Custom chandeliers formed from deer antlers contribute to the rural feel of the dining room, while fixtures in the main foyer and retail shop evoke images of the surrounding alpine forest. An exhibition kitchen fills one end of the dining space, trimmed with hammered steel in dark patinas. Here the club's chef can prepare meals at a peninsula of hardwood and stained concrete, in full view of his delighted guests.

位于科罗拉多州蒙特罗斯市西南方向32公里远处，这块24平方千米的土地周围满是野山花和成熟的山杨树，可将洛矶山脉的风景尽收眼底。这里，清新的山林空气中散发着野生鼠尾草的香味。奠基石高尔夫俱乐部的建筑正表达了这种山间小屋的感觉，体现了此地粗犷而朴实的美丽。这个坐落于山脚下的朴实建筑结构当初意在做为临时的建筑使用，但自从完工两年以来，其朴素的美丽给参观者们留下了深刻的印象：原木壁板和生锈的铁皮屋顶，屋顶四周是木制山墙和框架。碲化砂石制成的壁柱为建筑提供了坚实的基础，中央壁炉巨大的石烟囱从建筑的边线延伸出来。室内，灰白色硬木地板与深色的装饰和上方的木制横梁互相呼应。特制的鹿角形吊灯使餐厅有种乡村的感觉，而主门厅和零售商店内的布置则反映了周围高山森林的形象。餐厅一端有一个开放式厨房，由带铜绿的锻钢装饰而成。在这里，俱乐部的厨师可以在全是硬木和混凝土结构构成的厨房里为客人们准备餐点，在这里他们可以看到心情愉悦的客人们的用餐情况。

奠基石俱乐部附近还有一个小湖，当地丰富的野生动物经常光顾这个小湖。当俱乐部的会员们在露台上的火堆旁边小聚的时候，他们经常能看到只有大自然中才有的美景；这种美景会随着时段和季节的改变而发生变化。设计这座小小俱乐部建筑是为了与大背景相融合，而非与其相冲突。这样的设计与"大室外"无与伦比的美丽和谐统一。

1. Log siding and rusted steel roofing give the clubhouse its rustic mountain character
2. Chandeliers hung above dining tables are fashioned from deer antlers
3. An inviting stone fireplace occupies the centre of the Entry Hall
4. Hand-scraped recycled wood floors and timber beams make the Dining Room a natural extension of the alpine environment
5. Darkened copper wraps the hood and equipment in the Exhibition Kitchen
6. Bands of windows set between native sandstone piers offer views of the surrounding aspen forest

1.原木壁板和生锈的铁皮屋顶使俱乐部带有乡村风韵
2.餐桌上方悬挂的吊灯是鹿角造型
3.入口大厅的中央是石壁炉
4.手刮的可再生木地板和木横梁使餐厅成为高山环境的自然扩展
5.深色的铜质材料包裹着展示餐厅的各种器具
6.当地沙石制成的角柱中间有许多窗户，让客人们可以看到周围森林的景色

The Cornerstone clubhouse happily shares its grounds with a small lake that is frequented by the region's abundant wildlife. As club members gather at the fire pit on the covered patio they are often treated to majestic scenes that only nature can provide, scenes that can change with each passing hour of the day, and each season of the year. The small clubhouse building was designed to suit this setting, not confront it, and in doing so be harmonious with the incomparable beauty of the "great outdoors".

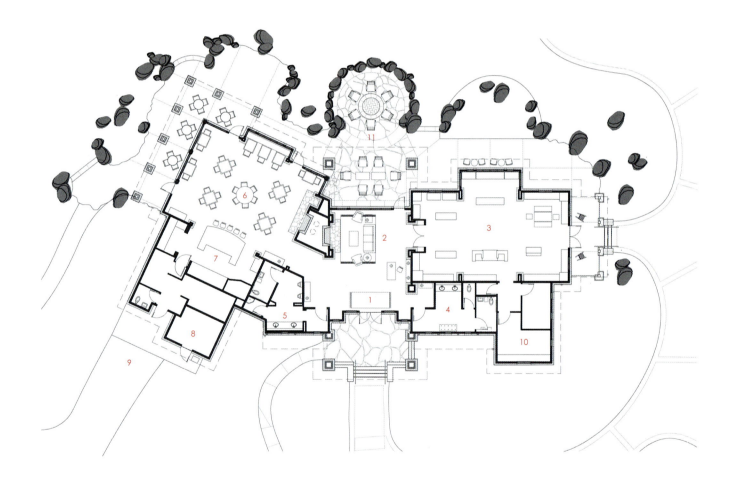

1. Entrance
2. Living room
3. Golf shop
4. Women's restroom
5. Men's restroom
6. Dining room
7. Exhibition kitchen
8. Back of house
9. Service drive
10. Offices
11. Patio

1.入口
2.起居室
3.高尔夫用品商店
4.女卫生间
5.男卫生间
6.餐厅
7.展示厨房
8.后勤区
9.服务车道
10.办公室
11.露台

3

Pine Canyon
松树谷俱乐部

Location:
Flagstaff, Arizona, USA

Designer:
Marsh & Associates, Inc.

Photographer:
Peter Wong, Vickie Atkins

Area:
3,530 m²

Completion date:
2007

项目地点：
美国，亚利桑那州，弗拉格斯塔夫市

设计师：
马什联合设计有限公司

摄影师：
王彼得，维基·阿特金斯

面积：
3530平方米

完成时间：
2007

In the pastoral national forest that surrounds the affable mountain town of Flagstaff, Arizona, the Pine Canyon clubhouse rests fittingly amidst the towering Ponderosa pines. Pine Canyon's bucolic two-storey golf clubhouse was designed as a series of interconnected structures, and its masses were arranged so as not to disturb the prevailing old-growth evergreens. The building features an exterior of multi-hued Telluride sandstone, split-log clapboards, and slate and rusted-steel roofs with copper accents. A series of peaked roof lines shed high-desert snowfalls, with muscular pine timber trusses for support. The building entrance features an elegant bridge over a cascading stream, whose flow meanders through outdoor terraces on its way to the adjacent lake. Stout hand-carved wooden doors greet visitors as they enter; inside, high vaulted ceilings, heavy timber trusses, and an immense four-sided fireplace beckon guests to mingle and relax in the central lodge. Custom lighting fixtures that evoke pine cone-laden branches are among the meticulous rustic mountain forest details. Pine Canyon also has a dedicated spa and fitness centre, whose lodge motif echoes the main clubhouse.

In 2007 the Club at Pine Canyon received the Gold Nugget Grand Award for Best Public/Private Recreational Facility from the Pacific Coast Builders Conference, and was named Best Private Clubhouse by Golf Inc. Magazine.

位于环绕弗拉格斯塔夫这座山城周围的国家森林中，松树谷会所处于美国黄松木的掩映之中。松树谷高尔夫俱乐部位于两层田园风格的建筑中，其各个建筑结构互相连接在一起，这样的设计安排是为了不破坏生长茂盛的成熟长青树。建筑的外部由多彩的碲化砂石、原木墙板和铜色锈铁屋顶构成。屋顶上是一系列由结实的松木构成的尖顶棚。入口处有一座精致的小桥，下面是潺潺的溪水，蜿蜒地流过室外露台流向相邻的湖泊。结实的手刻木门欢迎着来访的客人；里面是高高的拱形天花板、厚重的木架和一个巨大的四面壁炉，使客人们可以在这个乡村小舍中放松休息。仿照满是锥形叶子的松枝而特制的照明设施是乡村山林风格设计中的一个典型例子。松树谷里也有一个专门的SPA和健身中心，其室内的装饰风格与会所主建筑保持一致。

2007年，松树谷的这间俱乐部在太平洋海岸建筑师会议上获得了最佳公共/私人休闲设施金奖，并被高尔夫俱乐部有限公司杂志称为最佳私人会所。

1. The lodge-style Pine Canyon golf clubhouse is nestled amidst mature evergreens
2. Timber framing is plentiful in the clubhouse, echoing the high desert forest outside
3. The Wine Cave stylishly stores a broad selection of wines under perfect conditions
4. Timber trusses and warm colours bring a pastoral comfort to the Dining Room
5. The Men's Lounge features a fireplace constructed of native sandstone, and a sliding wood-and-glass wall that opens to connect the entire room with the outdoors
6. The Women's Lounge has clusters of soft seating
7. The Lounge Bar, a hub of social activity, has tin ceilings, soft booth seating, and plenty of bar stools

1.乡村小屋风格的松树谷高尔夫俱乐部坐落于四季常青的树木之间
2.俱乐部里采用了很多木制结构，与室外沙漠森林互相呼应
3.酒窖里存储了很多种保存良好的酒
4.木制结构和温暖的色调给餐厅带来了舒适的田园情趣
5.男士休闲吧里有用当地沙石建造的壁炉和可以滑动的木材和玻璃制成的墙，将室内和室外连接起来
6.女士休闲吧里有许多柔软的座椅
7.休闲酒廊是社交活动的中心，有锡制天花板、柔软的座椅和许多吧凳

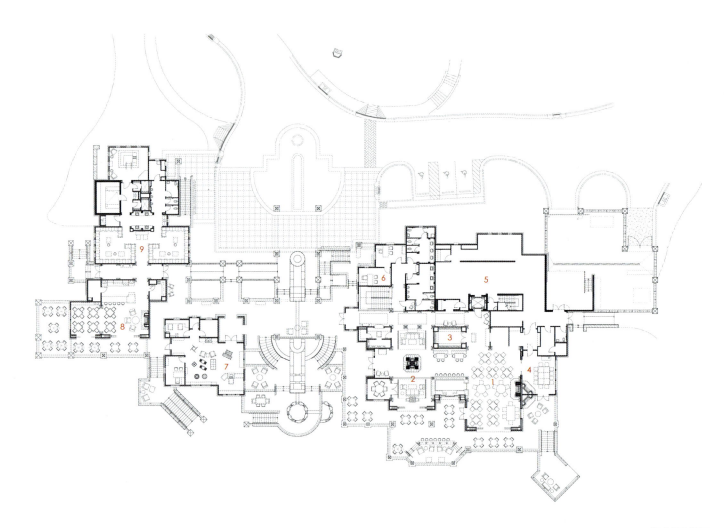

1. Mixed grill
2. Living room
3. Wine room
4. Private dining
5. Kitchen
6. General manager
7. Golf shop
8. Men's lounge
9. Men's lockers

1.混合烧烤餐厅
2.起居室
3.酒室
4.餐厅包间
5.厨房
6.总经理办公室
7.高尔夫用品商店
8.男士休闲吧
9.男士衣帽间

6

7

Pronghorn

叉角羚俱乐部

Location:
Bend, Oregon, USA

Designer:
Marsh & Associates, Inc.

Photographer:
Allen Kennedy, Mark Knight

Area:
5,020 m²

Completion date:
2007

项目地点：
美国，俄勒冈州，本德

设计师：
马什联合设计有限公司

摄影师：
艾伦·肯尼迪，马克·奈特

面积：
5020平方米

完成时间：
2007

Set in the dramatic high desert landscape of central Oregon, Pronghorn's three-storey clubhouse and spa makes the most of spectacular golf and mountain views. Expansive terraces grow out of the indigenous volcanic rock formations that encompass the back of the building, providing numerous intimate outdoor spaces. Borrowing materials from the land, the clubhouse is comprised of volcanic stone, natural stone veneers, heavy timbers, weathered stucco and antiqued copper windows. The use of natural materials consistent with the native landscape provides the historical ties to the community desired by the owner. Details such as signature timber trusses, custom designed to emulate pronghorn antelope antlers, and hand-crafted igneous rock terraces and tunnels inspired by the lava tubes found throughout the site, give this destination resort a character of all its own. Inside, wrought-iron light fixtures enhance tumbled travertine and rustic wood flooring. The carved furniture provides a traditional canvas for the rich colour palette of reds, greens, and browns, with distressed leathers and hair-on-hide accents.

This facility is unique in that it serves two world-class 18-hole golf courses with varying membership structures. The design challenge was to discreetly provide separation of the two levels of memberships, while also creating multiple co-mingling areas. One wing was

叉角羚俱乐部坐落于俄勒冈州中心的沙漠景观带中，这座三层楼高的会所和水疗养生中心，坐拥最壮观的高尔夫球场和山脉景观。广阔的露台从包围建筑后部的天然火山岩石结构中延伸出来，为顾客提供了许多隐秘的室外空间。俱乐部利用得天独厚的自然资源，由火山岩石、天然石饰面、厚重的木料、风化灰泥和古老的铜窗户构成。自然材料与当地景观相一致，达到了主人将社区与历史相联系的愿望。细节方面如专门模仿叉角羚角设计而成的特色木制结构，受随处可见的熔岩管的启发，用手工铺成的火成岩露台和过道等都是这个度假胜地的特色。室内，熟铁制成的照明装置为乡村风情的石灰华和木制地板增添了生趣。雕花的家具为室内丰富的红、绿、棕色调色板提供了画布，室内还配有皮革等饰物。

这座建筑是独特的，内含两个世界级的18洞高尔夫球场，有不同的会员结构。设计所受到的挑战便是要将两个不同等级的会员结构分开，同时又创建出各种各样的混合区域。一边配有精致的陈设和装饰，另一边则以乡村粗犷的特征为主题。两边都为顾客提供了舒适、愉悦、独特的享受，这与主人的预想相一致。公共区域将不同的装饰细节融合在一起，既为顾客提供了隐秘的交流场所，也为其提供了广阔的聚会场所。因此，不同级别的会员和他们的客人们，无论身处哪个区域，都可以体验到这里田园般的环境所提供的丰富空间与秀丽景色。

除了俱乐部主室，马什联合设计有限公司还设计了一系列的小茅屋，与主室的设计主题相似，客人们可以在这里住宿。

1. The Pronghorn Club occupies a dramatic setting in the high desert of Central Oregon
2. French doors in the private "Solstice Room" open to a balcony with a second-storey view over the golf course
3. Soft leather-clad furniture in the Lounge offers a variety of opportunities for socialising, both indoors and outdoor
4. Multiple terraces, patios, and balconies draw guests out to enjoy the cool evening air
5. The main Dining Room features craftsman-style windows and custom light fixtures, and earth-tone carpet and fabrics
6. A large stone fireplace opens to both sides of the Main Hall

1.叉角羚俱乐部位于俄勒冈中央沙漠之中，地理位置非常引人注目
2.二楼"顶点"私人包房装有法式门，通向阳台，可以俯瞰高尔夫球场
3.休闲吧内柔软的皮质家具为顾客提供了一系列进行室内外社交活动的机会
4.各种各样的露台和阳台吸引顾客来到室外享受夜晚清凉的空气
5.主餐厅内有手工艺风格的窗户和专门制作的灯具，以及泥土格调的地毯和织物
6.主厅内一个大型石材壁炉向两边开启

articulated with more refined finishes and details, while the other embraces the rustic character of the locale. Both provide comfort, amenities and exclusivity consistent with the owner's vision. The common areas incorporate a marriage of the disparate details while providing both intimate enclaves and grand gathering spaces. As a result, members at each level and their guests, separately or together, can experience spaces and vistas with a richness borne of this idyllic setting.

In addition to the main clubhouse, MAI also designed a series of cottages, which feature a similar design theme and allow for on-site accommodations at the resort.

1. Men's lounge
2. Women's lounge
3. Dining
4. Private dining
5. Kitchen
6. Bar
7. Lobby
8. Administration
9. Women's VIP lockers
10. Fitness
11. Men's VIP lockers
12. Golf shop
13. Spa/Treatment

1.男士休闲吧
2.女士休闲吧
3.餐厅
4.餐厅包间
5.厨房
6.酒吧
7.前厅
8.行政办公室
9.女士贵宾储物室
10.健身房
11.男士贵宾储物室
12.高尔夫用品商店
13.水疗/理疗室

Social Hollywood

好莱坞社交俱乐部

Location:
Los Angeles, CA, USA

Designer:
Mark Zeff

Area:
2,508 m^2

Completion date:
2006

项目地点：
美国，加利福尼亚州，洛杉矶市

设计师：
马克·泽夫

面积：
2508平方米

完成时间：
2006

Social Hollywood, in Los Angeles, California, is a multifaceted restaurant and lounge. The Hollywood athletic club was once known as the "playground to the stars". The private club attracted the biggest names of the golden era, including its founders Cecil B Deville, Charlie Chaplin, Lon Chaney and Rudolph valentine. Zeff's creative design vision was to acknowledge the building's history and the glamour of old Hollywood, and to blend it with modern international design.

A 2,508-square-metre space that includes a restaurant, bar and private membership area housed in the former Hollywood Athletic Club. Zeff converted the 1923 landmark playground of the stars into a modern-day Casablanca. Social Hollywood has a Moroccan motif and a sense of playfulness that flows through the restaurant. Zeff set the tone with furnishings and objects purchased on trips to Marrakech, Tangier and Fez. Stunning juxtapositions in the dining room include intricately carved and whitewashed Moroccan wood chairs set against a background that features shifting moving projected animations. Social Hollywood has the aura of the Art Deco golden age feel, updated with a modern sexy vibe. For the members-only second floor Zeff gave each space an individual feel, the game room with billiards and play stations has ebony stained woven-wood walls, and a brighter palette defines the green room lounge.

好莱坞社交俱乐部位于加利福尼亚州洛杉矶市，是一间多功能餐厅和休闲吧。好莱坞运动俱乐部曾被誉为"明星们的游乐场"。这间私人俱乐部吸引了黄金时代最有名气的人们，包括它的创始人塞西尔·B·德维尔、查理·卓别林、朗·钱尼和鲁道夫·瓦伦丁。泽夫设计的创造性视野就在于保留了这座建筑的历史性和旧好莱坞的光辉，将其与现代国际化设计相融合。

这个2508平方米的空间包含了一间餐厅、一间酒吧和一处私人会所，位于前好莱坞运动俱乐部的旧址上。泽夫将这个1923年典型的明星娱乐场所转化成了一座现代的"卡萨布兰卡"。好莱坞社交俱乐部以摩洛哥风格为主题，嬉戏幽默的韵味弥漫在整个餐厅氛围里。泽夫通过使用家具和陈设为这里定下了基调。这些家具和陈设购买于去马拉喀什、丹吉尔和非斯旅行的途中。餐厅内令人惊叹的配置包括精雕细琢粉刷雪白的摩洛哥木椅，背景是预置的动画画面。好莱坞社交俱乐部内有种黄金时代艺术装饰的味道，配合以现代的引人入胜的意境。对于会员专用的二楼，泽夫为每个独立的空间设计了不同的感觉。桌球室与游戏室里有黑檀木墙，而休息室则是以绿色等鲜亮颜色为主色调。

1. The white exterior wall evokes the interior luxury
2. The lighting creates a bright and warm space
3. The sofas in the "Green Room" provides comfort seating
4. The antiques shining with candlelight
5. The ceiling, wine shelf and the decorations combining together to create a luxury space
6. The candlelights shining with the chandeliers above

1.白色的外墙预示着室内的豪华
2.灯光营造了明亮温暖的空间
3.绿室内的沙发为顾客提供了舒适的享受
4.各种古董在烛光照耀下散发着光芒
5.天花板、酒架和装饰共同创造了奢华的空间
6.烛光与上方的吊灯交相辉映

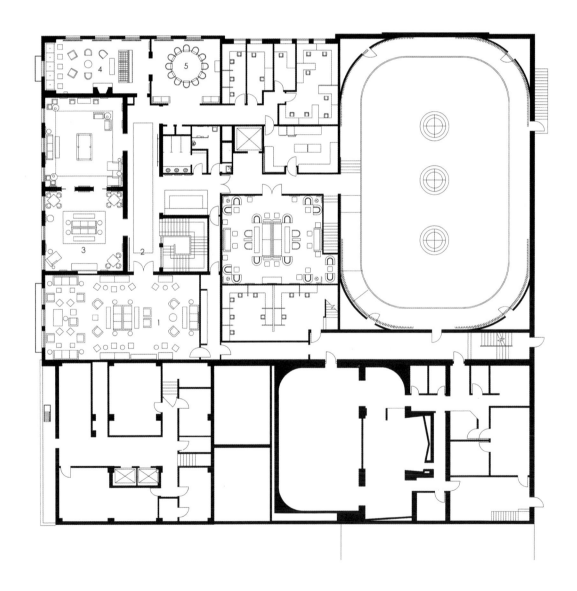

1. Velvet room   1.天鹅绒室
2. Second floor hallway   2.二楼走廊
3. Gaming room   3.游戏室
4. Green room   4.绿室
5. Private dining room   5.餐厅包间

4

5

6

1. The bar with great elegance
2. The informal dining provide a casual atmosphere
3. The lounge with comfortable seating
4. The screening room with special ceiling design
5. The facilities in the fitness centre
6. Guests could enjoy their time in the library
7. The neat salon

1.典雅的酒吧
2.非正式的餐厅为顾客提供了随意的就餐环境
3.休息室内有舒适的座椅
4.放映厅中的天花板设计独特
5.健身中心内的健身器材
6.客人们可以在图书室中享受美好时光
7.整洁的沙龙

elegantly used throughout the floor, create a monolithic sense of space. At the top floor the spa's individual treatment suites are the most dramatic incarnation of the manipulated wood material. Critical surfaces become those one focuses on during a treatment – the ceiling and the floor. Parquet wood flooring is rendered in a custom triangulated pattern reminiscent of the chamfer as if it were flattened. Treatment cabinets and ceilings are an exploded version of the chamfer and give way to an oculus of light which defines each room as a core space organised around each individual member's needs.

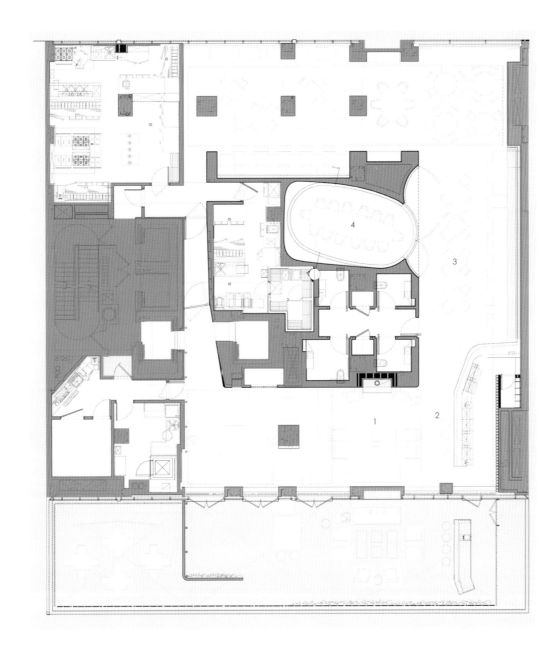

1. Lounge	1.休闲吧
2. Bar	2.酒吧
3. Informal dining	3.非正式餐厅
4. Private dining	4.餐厅包房

3

4

5

1. The entrance invites you in with a series of red lacquered gates
2. The gaming hall provides guests with great pleasure
3. The ceiling shining with red glowing light
4. The comfortable lounge
5. The carpet of black-and-grey line imitates the run way of planes
6. The overall design of the gaming hall is luxury
7. The walls are constantly adorned with yellow and black painted glass and stripes of gold glass

1.入口处的红漆门欢迎人们的光临
2.博彩大厅为顾客提供了极大的乐趣
3.天花板散发着红光
4.舒适的休息室
5.黑灰相间的地毯是模仿飞机的跑道设计而成的
6.博彩大厅的整体设计都很豪华
7.墙面装饰有黄黑漆玻璃和金色玻璃条

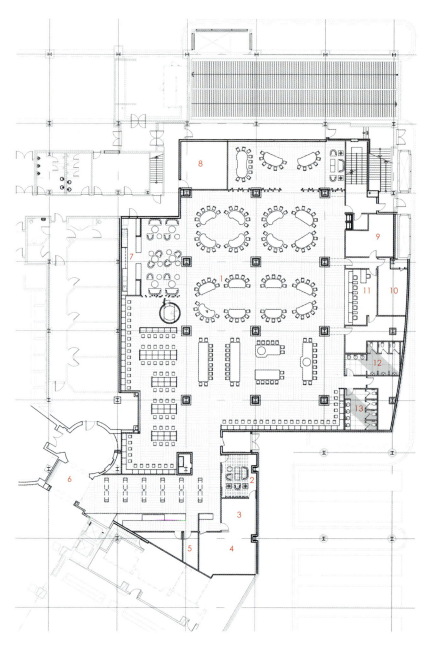

1. Casino Hall   1.博彩大厅
2. VIP room   2.贵宾室
3. Office   3.办公室
4. Service station   4.服务站
5. Back office /check-in   5.办公室/登记处
6. Entrance   6.入口
7. Bar   7.酒吧
8. Kitchen   8.厨房
9. Bank   9.银行
10. Count room   10.会计室
11. Cash exchange   11.现金兑换处
12. Toilet (M)   12.男卫生间
13. Toilet (W)   13.女卫生间

Walker Hill Paradise Casino

华克山庄天堂博彩俱乐部

Location:
Seoul, Korea

Designer:
Studio GAIA Architecture and Interiors

Photographer:
Studio GAIA Architecture and Interiors

Completion date:
2006

项目地点：
韩国，首尔

设计师：
盖亚建筑与室内设计工作室

摄影师：
盖亚建筑与室内设计工作室

完成时间：
2006

The Walker Hill Paradise Casino is located in Seoul, Korea. The reoccurring theme in the design is nature and lighting is key in creating this atmosphere for the casino. Waterfalls encased in clear glass help open up the space while creating sections within the casino. Throughout the casino onyx walls are lit up from within to help accent a warm natural lighting. The main room serves as the most striking element as the ceiling is made up of ever-changing new matt-lit ceilings. The carpet chosen is a pattern of red roses that help give the space a complimentary area when reflected with the red tinted ceiling. In order to keep the flow of the casino, some of the walls are black reflected surfaces. This casino is an innovative solution that encompasses elements of nature creating a strong reaction in every detail and facet.

华克山庄天堂博彩俱乐部位于韩国首尔。设计所要强调的主题是"自然"，因此灯光设计对于营造赌场的自然氛围非常关键。透明玻璃内的瀑布将室内的空间拓宽，创建出不同的赌场分区。赌场内所有缟玛瑙墙面都从内部发出柔和的光，给人以温暖、自然的感觉。主室是最引人注目的，因为采用了千变万化的新式亚光天花板。红玫瑰地毯与彩色天花板交相辉映，更为空间增添了色彩。为了反映出俱乐部内的客流状况，一些墙采用了黑色反光墙面。这间俱乐部的设计打破常规，每一个细节、每一个层面都强烈体现了自然的元素。

1. The decoration on the wall is very exclusive
2. The carpet with red-rose patterns
3. The gaming hall
4. The onyx wall is lit up from within with warm light
5. The comfortable seating combines with the ceiling
6. Lighting plays an important role in creating a natural atmosphere
7. The carpet reflects with the red tinted ceiling

1.墙上的装饰非常独特
2.地毯上有红玫瑰图案
3.博彩大厅
4.缟玛瑙墙面由内而外散发出温暖的光
5.舒适的座椅与天花板相呼应
6.灯光在营造自然氛围上发挥了重要作用
7.地毯与红色天花板交相辉映

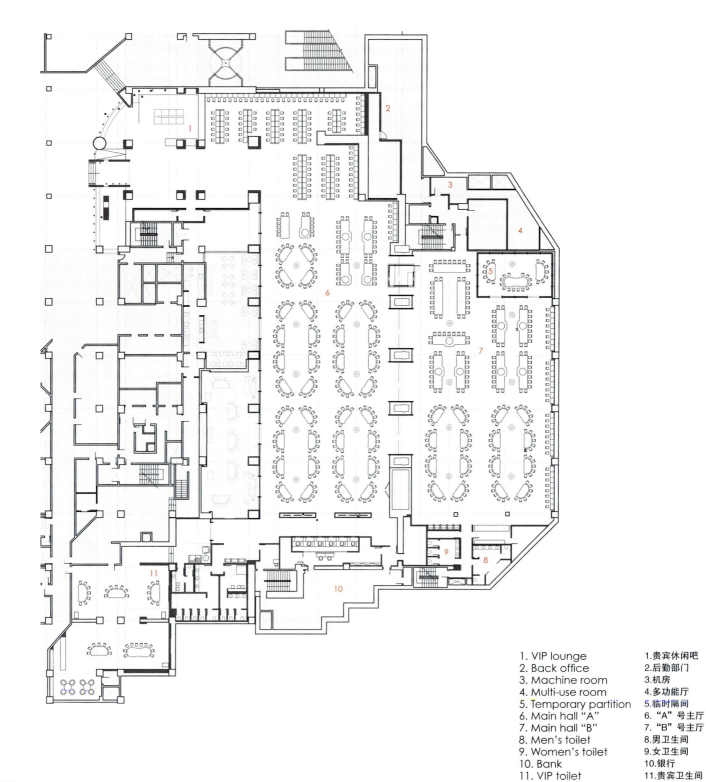

1. VIP lounge
2. Back office
3. Machine room
4. Multi-use room
5. Temporary partition
6. Main hall "A"
7. Main hall "B"
8. Men's toilet
9. Women's toilet
10. Bank
11. VIP toilet

1.贵宾休闲吧
2.后勤部门
3.机房
4.多功能厅
5.临时隔间
6."A"号主厅
7."B"号主厅
8.男卫生间
9.女卫生间
10.银行
11.贵宾卫生间